Cambridge Elements

Elements in Geochemical Tracers in Earth System Science
edited by
Timothy Lyons
University of California
Alexandra Turchyn
University of Cambridge
Chris Reinhard
Georgia Institute of Technology

LOCAL AND GLOBAL CONTROLS ON CARBON ISOTOPE CHEMOSTRATIGRAPHY

Anne-Sofie Ahm
Princeton University and University of Victoria
Jon Husson
University of Victoria

CAMBRIDGE
UNIVERSITY PRESS

University Printing House, Cambridge CB2 8BS, United Kingdom

One Liberty Plaza, 20th Floor, New York, NY 10006, USA

477 Williamstown Road, Port Melbourne, VIC 3207, Australia

314–321, 3rd Floor, Plot 3, Splendor Forum, Jasola District Centre,
New Delhi – 110025, India

103 Penang Road, #05–06/07, Visioncrest Commercial, Singapore 238467

Cambridge University Press is part of the University of Cambridge.

It furthers the University's mission by disseminating knowledge in the pursuit of education, learning, and research at the highest international levels of excellence.

www.cambridge.org
Information on this title: www.cambridge.org/9781009013956
DOI: 10.1017/9781009028882

First published 2022

A catalogue record for this publication is available from the British Library.

ISBN 978-1-009-01395-6 Paperback
ISSN 2515-7027 (online)
ISSN 2515-6454 (print)

Local and Global Controls on Carbon Isotope Chemostratigraphy

Elements in Geochemical Tracers in Earth System Science

DOI: 10.1017/9781009028882
First published online: March 2022

Anne-Sofie Ahm
Princeton University and University of Victoria

Jon Husson
University of Victoria

Author for correspondence: Anne-Sofie Ahm, annesofieahm@uvic.ca

Abstract: Over million-year timescales, the geologic cycling of carbon controls long-term climate and the oxidation of the Earth's surface. Inferences about the carbon cycle can be made from time series of carbon isotopic ratios measured from sedimentary rocks. The foundational assumption for carbon isotope chemostratigraphy is that carbon isotope values reflect dissolved inorganic carbon in a well-mixed ocean in equilibrium with the atmosphere. However, when applied to shallow-water platform environments, where most ancient carbonates preserved in the geological record formed, recent research has documented the importance of considering both local variability in surface water chemistry and diagenesis. These findings demonstrate that carbon isotope chemostratigraphy of platform carbonate rarely represents the average carbonate sink or directly records changes in the composition of global seawater. Understanding what causes local variability in shallow-water settings, and what this variability might reveal about global boundary conditions, are vital questions for the next generation of carbon isotope chemostratigraphers.

Keywords: carbon isotopes, global carbon cycle, platform carbonates, carbonate diagenesis, chemostratigraphy

ISBNs: 9781009013956 (PB), 9781009028882 (OC)
ISSNs: 2515-7027 (online), 2515-6454 (print)

Contents

1 Introduction

The geologic carbon cycle is central to our understanding of the evolving habitability of planet Earth. The solid Earth outgasses carbon as CO_2 to the ocean-atmosphere system, and sedimentary basins bury carbon as either carbonate minerals (calcite, $CaCO_3$, or dolomite, $CaMg(CO_3)_2$) or organic matter (CH_2O). The burial of carbonate is a product of chemical weathering of igneous minerals, which generates the necessary alkalinity for carbonate mineral precipitation from seawater. Owing to a temperature dependence of chemical reaction rates, chemical weathering (and associated carbonate burial) acts as a planetary thermostat, regulating the greenhouse gas CO_2 and stabilizing global temperatures on long timescales (>10^5 years; Walker et al. 1981). By contrast, organic matter formation is the result of biological activity. If the product of oxygenic photosynthesis is $CO_2 + H_2O \Leftrightarrow CH_2O + O_2$, the burial of organic carbon results in the net release of free O_2 to the surface environment.

Sedimentary burial fluxes of carbon are connected both to the long-term maintenance of an equable climate (e.g., Walker et al. 1981) and the oxygenation of the surface of the Earth (Broecker 1970). However, direct constraints on carbon burial fluxes, and their use to study Earth history, are rare, owing both to the difficulties in building geologic syntheses of sedimentary volumes (Ronov et al. 1980) and to the uncertainties surrounding how erosion and rock cycling have affected such records (Gregor 1970). As a result, the measurement of proxies is the dominant approach for the study of the ancient carbon cycle – specifically, the measurement of the ratio of stable carbon isotopes (^{12}C and ^{13}C) in carbonate rocks and sedimentary organic matter. Under specific assumptions about both how carbon behaves in the ocean-atmosphere system and how surface geochemistry is recorded in the sedimentary record, inferences about the geologic carbon cycle can be made from time series of carbon isotope ratios. These inferences include the origin of life (Schidlowski et al. 1975), transient increases in atmospheric CO_2 (e.g., Berner 2006), and the burial history of organic carbon across the Phanerzoic (Broecker 1970) and Precambrian (Knoll et al. 1986). By direct consequence of this interpretative framework, carbon isotopic values can be used as tools of stratigraphic correlation. Namely, on timescales of sedimentary rock formation, perturbations and changes to the carbon cycle should be recorded in globally disparate basins as synchronous events, and thus are useful as chronostratigraphic markers for intra-basinal and interbasinal correlation models. This application is referred to as *carbon isotope chemostratigraphy*, and is relied on heavily for the study of the Precambrian when index fossils needed for biostratigraphy are lacking (Knoll et al. 1986).

In this Element, we highlight the developments, potential pitfalls, and future potentials of carbon isotope chemostratigraphy. As we explain later, the validity of its application is dependent on assumptions about the global carbon cycle, and about how the isotope geochemistry of sedimentary carbonates and organic matter records conditions of the surface environment from which they formed. Of singular importance is the observation that modern shallow-water depositional systems are dominated by local carbon cycling, leading to large differences between the carbon isotope composition of modern, shallow-water $CaCO_3$ sediment and average carbonate burial. Understanding this disconnect is important, because all sediments older than ~200 million years old formed in analogous environments, as abyssal sediments are recycled and destroyed due to seafloor spreading and subduction. Disentangling the mixture of global and local control in modern carbon isotope values is therefore vital for the interpretation of deep time records, and for the basis of carbon isotope chemostratigraphy.

2 Systematics of Carbon Isotope Chemostratigraphy

2.1 Development and History

The groundwork for carbon isotope chemostratigraphy was laid by the pioneering research in isotope geochemistry by Alfred Nier (Nier and Gulbransen 1939) and Harold Urey (Urey et al. 1936, Urey 1947). Both made the earliest measurements of the ratios of carbon isotopes in natural materials by mass spectrometry (Figure 1A). By convention, rather than discussing such values as simple ratios (e.g., $^{13}R = {}^{13}C/{}^{12}C$), carbon isotopic values are expressed in the δ-notation relative to a common standard (V-PDB, Craig 1953):

$$\delta^{13}C = \left(\frac{^{13}C_{sample}}{^{12}C_{sample}} - 1\right) * 1000. \qquad (1)$$

With regards to chemostratigraphy, one of the most important early findings was that organic matter is depleted in ^{13}C relative to carbonate minerals (Nier and Gulbransen 1939). This discovery forms the backbone of all studies using $\delta^{13}C$ values to study the carbon cycle in deep time. The first long-term record of Phanerozoic carbonate $\delta^{13}C$ values was published in 1964 (Keith and Weber 1964), followed by the first comprehensive Precambrian record in 1975 (Schidlowski et al. 1975), both records remarking on the apparent lack of long-term variability in $\delta^{13}C$ values. These observations led to the prevailing hypothesis that the burial of organic carbon (and, consequentially, atmospheric oxygen levels) have been relatively stable throughout the Phanerozoic (Broecker 1970).

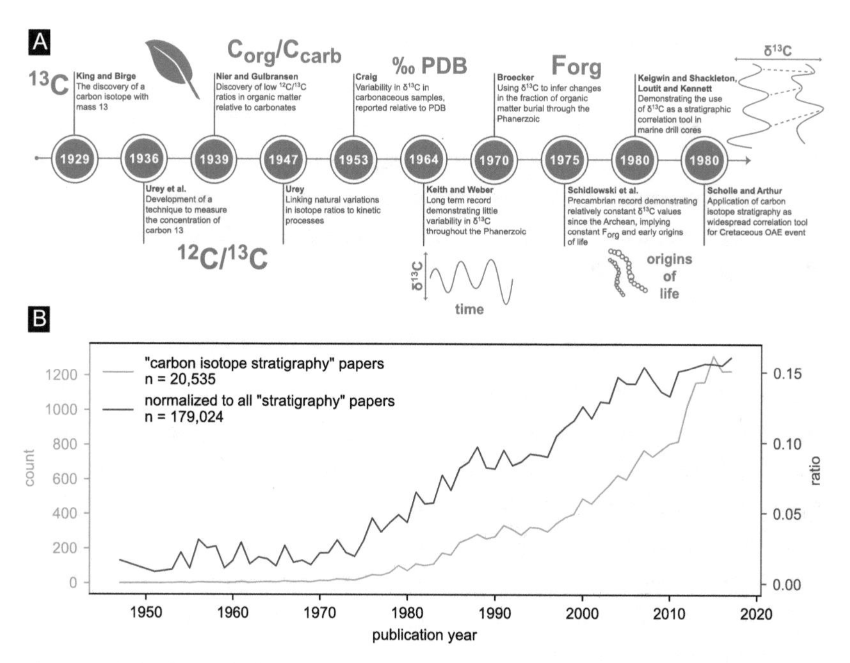

Figure 1 (A) The early history of carbon isotope research, from the discovery of $\delta^{13}C$ in 1929 (King and Birge 1929) to 1980, listing some of the most consequential papers that underpin the development of carbon isotope chemostratigraphy. (B) The number of peer-reviewed papers published per year that contain the phrases "carbon isotopes" and "stratigraphy" (blue, left y-axis), extracted from the GeoDeepDive digital library (https://geodeepdive.org). This record is also normalized to the yearly count of papers that contain the word "stratigraphy" (red, right y-axis), and shows that the acceleration in rate of published $\delta^{13}C$ chemostratigraphy papers is not driven solely by overall growth in scientific research output. In both records, an uptick in the early 1970s is apparent, coincident with the initiation of the Deep Sea Drilling Program Data in 1968.

The later development of $\delta^{13}C$ values as a tool for stratigraphic correlation is linked closely to the initiation of the Deep Sea Drilling Program (1968–83; Figure 1A). Following the success of using $\delta^{18}O$ for correlation in Pleistocene and Pliocene deep sea sediment cores, carbon isotope stratigraphy was first used in the late 1970s (Figure 1B), based on the observation of reproducible and coherent changes in $\delta^{13}C$ values of carbonate (excursions of $\sim -0.5‰$) in Miocene sediments across the Pacific Ocean (Loutit and Kennett 1979; Keigwin and Shackleton 1980). Contemporaneous with these developments, the observation of significant short-term fluctuations – excursions – of carbonate $\delta^{13}C$ values in Cretaceous sedimentary rocks, associated with ocean anoxic events (OAEs), pushed the use of carbon isotope stratigraphy into older portions of the geologic record (Scholle and Arthur 1980). Today, numerous records of $\delta^{13}C$ values from Cenozoic deep marine sediment cores have demonstrated systematic variations across the globe that correlate with major climatic events, used for global correlation and age model construction (e.g., Westerhold et al. 2020).

2.2 One-box Model of the Carbon Cycle

At its core, carbon isotope chemostratigraphy builds on the assumption that the $\delta^{13}C$ values of carbonate ($\delta^{13}C_{carb}$) reflect the $\delta^{13}C$ values of dissolved inorganic carbon (DIC) ($\delta^{13}C_{DIC}$; Figure 2) in a well-mixed ocean in equilibrium with the atmosphere (Kump and Arthur 1999). In this view, independent of geographic location, all carbonate precipitated and deposited has the same $\delta^{13}C$ value, and excursions in $\delta^{13}C_{carb}$ will be synchronous and useful as stratigraphic tie points between and within sedimentary basins.

The $\delta^{13}C_{DIC}$ value of a well-mixed ocean in equilibrium with the atmosphere – in this Element, referred to as "global $\delta^{13}C_{DIC}$" – is set by the relative sizes and isotopic values of carbon sources and sinks. The primary sources of carbon to the system are weathering (F_w), metamorphism, and volcanic outgassing (F_{volc}), while the sinks are burial of organic carbon ($F_{b,org}$) and carbonate ($F_{b,carb}$), with each flux carrying a (potentially) predictable isotopic value (Kump and Arthur 1999). Most commonly, the $\delta^{13}C$ value of CO_2 from volcanic and metamorphic outgassing is assumed to reflect the isotopic composition of the mantle (~ –5‰), the $\delta^{13}C_{carb}$ value is assumed to be equal to the composition of seawater, and the $\delta^{13}C_{org}$ value is assumed to be depleted in ^{13}C by approximately ~25‰ relative to $\delta^{13}C_{carb}$ (Hayes et al. 1999; Kump and Arthur 1999).

Based on the simple framework outlined refers to the previous paragraph, mathematical expressions can be written for carbon mass (M) and carbon isotope (δ_{DIC}) balance in the ocean-atmosphere system (Figure 2). The change in the mass of carbon in the system through time (dM/dt) equals the balance between sources and sinks:

$$\frac{dM}{dt} = F_w + F_{volc} - F_{b,org} - F_{b,carb}. \tag{2}$$

By the same logic, isotope mass balance can be calculated if each flux or reservoir is multiplied by its respective $\delta^{13}C$ value:

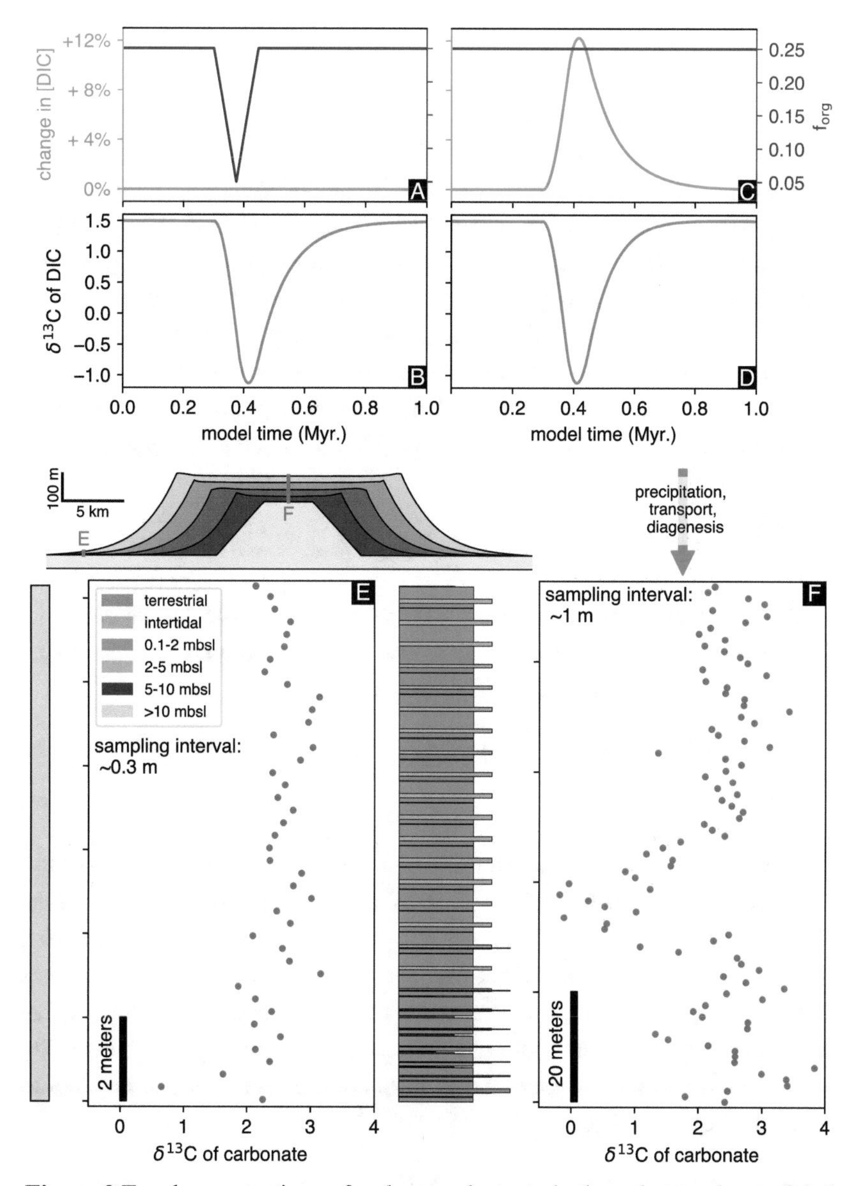

Figure 2 Toy demonstrations of carbon cycle perturbations that can be modeled using Equation 5 (see Subsection 2.2), and how these perturbations might be recorded in shallow water carbonate strata. A transient decrease in the relative burial flux of organic carbon (A) will result in a negative excursion in $\delta^{13}C$ values of DIC (B). If CO_2 input to the ocean atmosphere is increased via a pulse of oxidized organic carbon, the amount of DIC would increase (C) and an identical negative excursion can be produced (D), but with a very different driver than in (A) and (B). How these seawater signals are recorded in sediments

Caption for Figure 2 (cont.)

is explored with a simple numerical model of a shallow-water carbonate platform, with $CaCO_3$ sediment production on the platform edges (e.g., fringing reefs), which is then transported via hillslope diffusion. As the platform aggrades and progrades, the proscribed $\delta^{13}C_{DIC}$ signal from (B) or (D) is recorded with a fractionation of 1‰ and added "geological noise" (e.g., effects from differential mineralogy, organic matter respiration, etc.). In distal settings (E), where all sedimentation occurs in depths >10 meters below sea level (mbsl), accumulation rates are low and carbon cycle perturbations are captured poorly even with dense sampling. In platform interior settings (F), accumulation rates are higher, and the resulting $\delta^{13}C_{carb}$ record is expanded.

$$\frac{dM\delta_{DIC}}{dt} = \delta_w F_w + \delta_{volc} F_{volc} - \delta_{org} F_{b,org} - \delta_{carb} F_{b,carb}. \tag{3}$$

By rearranging these equations using the product rule of calculus ($dM\delta/dt = M * d\delta/dt + \delta * dM/dt$), it is possible to isolate $d\delta/dt$ and derive a time-dependent differential equation for ocean $\delta^{13}C_{DIC}$ values:

$$\frac{dM\delta_{DIC}}{dt} = \frac{F_w(\delta_w - \delta_{DIC}) + F_{volc}(\delta_{volc} - \delta_{DIC}) - F_{b,org}(\delta_{org} - \delta_{DIC}) - F_{b,carb}(\delta_{carb} - \delta_{DIC})}{M}. \tag{4}$$

This expression is typically simplified further by making three assumptions (Kump and Arthur 1999). First, the $\delta^{13}C$ value of carbonate is set equal to the value of DIC ($\delta_{DIC} = \delta_{carb}$). Second, the input fluxes from weathering and volcanic outgassing are merged into a single input flux from rivers (F_{riv}). Third, the $\delta^{13}C$ value of organic carbon is defined by an average offset from carbonate ($\Delta_B = \delta_{org} - \delta_{carb}$). We explore the validity of these assumptions in more detail in Section 3, but they can be used to simplify Equation 4 to:

$$\frac{d\delta_{carb}}{dt} = \frac{F_{riv}(\delta_{riv} - \delta_{carb}) - F_{b,org} * \Delta_B}{M}. \tag{5}$$

This box model approach (Equation 5) shows that the $\delta^{13}C$ value of the global ocean (and, by extension, that of buried marine carbonate) may evolve with time through numerous forcings, with different processes often generating the same δ_{carb} time series. For example, total carbon input (F_{riv}) and burial ($F_{b,carb} + F_{b,org}$) fluxes can remain equal, while the ratio of $F_{b,org}$ relative to $F_{b,carb}$ may decrease. In this scenario, M will remain the same but $\delta^{13}C$ of DIC will fall (Figure 2A–B). Transient imbalances in the input/output fluxes of carbon can lead to changes in both M and $\delta^{13}C_{DIC}$: oxidation of a large pool of organic carbon ($\delta^{13}C$~ −25‰) will increase ocean DIC concentration and will lower its $\delta^{13}C$ value (Figure 2C–D).

When considering timescales much longer than the residence time of carbon in the ocean-atmosphere system (>10^5 years), the carbon cycle box model is in a steady state, and the input and output mass fluxes must balance one another. In this scenario, Equation 5 can be further simplified by setting $d\delta_{carb}/d\delta_t = 0$ and setting the input flux of carbon equal to the burial fluxes ($F_{riv} = F_{b,org} + F_{b,carb}$):

$$\frac{F_{org}}{F_{org} + F_{carb}} = \frac{\delta_{riv} - \delta_{carb}}{\Delta_B} = f_{org}. \tag{6}$$

The burial fraction of organic carbon (F_{org}) relative to total carbon burial ($F_{org} + F_{carb}$) is commonly abbreviated in the literature as f_{org} (Figure 2A and C). Equation 6 demonstrates that if we know both the average isotope offset between organic matter and carbonate (e.g., $\Delta_B = -25‰$; Hayes et al. 1999) and the average isotopic value of the carbon input flux (δ_{riv}, assumed to equal the mantle value of ~−5‰), then it is possible to calculate f_{org} (Kump and Arthur 1999). By illustration, if δ_{carb} is set to the modern DIC value of ~0%, then f_{org} is 0.2, implying that on Earth today 20 percent of the carbon sink is organic carbon burial and 80 percent is carbonate burial.

The hypothesis that $\delta^{13}C_{carb}$ values can be directly linked to oxygen production by estimating the relative burial flux of organic carbon (f_{org}) is widely used to interpret deep time $\delta^{13}C_{carb}$ records, both in the Precambrian (e.g., Knoll et al. 1986; Canfield et al. 2020) and Phanerozoic (e.g., Broecker 1970; Saltzman 2005; Berner 2006). To apply this framework, it is assumed that $\delta^{13}C_{carb}$ values measured across several stratigraphic columns are representative of the average carbonate burial sink and that stratigraphic trends in $\delta^{13}C_{carb}$ represent secular changes in the global carbon cycle. Later, we examine this assumption more closely. Specifically, we investigate the effects of local isotopic variability on stratigraphic records of $\delta^{13}C_{carb}$ and the implications for global mass balance.

3 Local Controls and Issues of Fidelity and Diagenesis

In theory, according to the model framework developed above (Section 2.2), time series of $\delta^{13}C$ values of either carbonate or organic carbon can be used to reconstruct global ocean $\delta^{13}C_{DIC}$ values. In practice, $\delta^{13}C$ values of carbonate ($\delta^{13}C_{carb}$) are far more commonly measured, owing in large part to the high throughput capabilities of sample preparation and modern instrumentation. When interpreting $\delta^{13}C$ records in terms of the global carbon cycle (e.g., Equations 5 and 6), the important question is: Does the record capture a representative $\delta^{13}C$ value of average carbonate burial? For Cenozoic deep-marine $\delta^{13}C$ records, datasets can be collected that cover large geographical areas of the deep ocean seafloor and, when stacked in time bins, are likely to satisfy this constraint (e.g., Westerhold et al. 2020). By contrast, the chemostratigraphic records from shallow-water settings, such as carbonate platforms or epeiric seas, are complicated by a much larger range in $\delta^{13}C_{carb}$ values owing to local carbon cycling (Figure 3; Holmden et al. 1998; Panchuk et al. 2006).

Today, it is accepted that the average $\delta^{13}C_{carb}$ value of carbonate forming on shallow-water platforms does not represent the $\delta^{13}C_{carb}$ value of average carbonate burial, even when integrated across the globe (Swart 2008). Prior to the evolution of a deep-marine carbonate sink in the mid-Mesozoic, carbonate platforms likely played a larger role in the global carbonate burial budget (Opdyke and Wilkinson 1988), meaning that average platform $\delta^{13}C_{carb}$ would have been closer (or equal) to globally average carbonate $\delta^{13}C$. As a result,

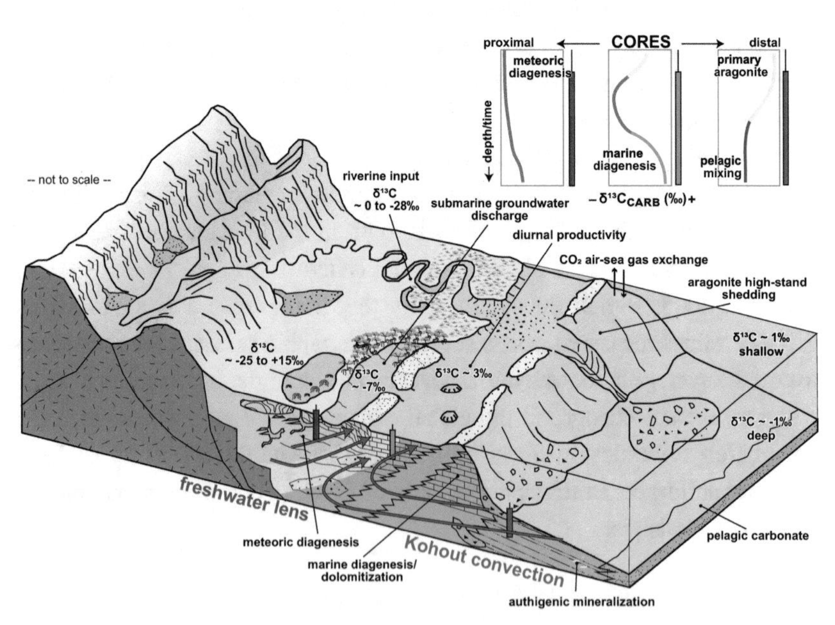

Figure 3 Schematic illustrating the variability in surface water $\delta^{13}C$ across different depositional environments in carbonate platform or epeiric sea settings (see text in Section 3 for a description of each environment). Values of $\delta^{13}C$ refer to DIC, unless stated otherwise. Core profiles illustrate the coincident variability in bulk carbonate $\delta^{13}C$ that can be expected as products of local surface water chemistry and carbonate diagenesis.

following the approach from the Cenozoic, a hypothesis for the pre-Mesozoic is that changes in the global carbon cycle can be evaluated by collecting correlative chemostratigraphic records of $\delta^{13}C$ values across several continents. However, even when independent tools such as biostratigraphy or geochronology are available for correlation, perturbations to the global carbon cycle that occur on timescales of <100,000 years (Figure 2A–D) are challenging to resolve for even the best chronostratigraphic age models (Schoene 2014; Holland 2020).

Given the short residence time of carbon (~10^5 years), if shallow-water carbon successions are only broadly correlative (e.g., at the 0.5–5 million year level), or only recorded in a subset of basins globally, excursions in $\delta^{13}C_{carb}$ values could easily represent local processes rather than global perturbations while satisfying global mass balance (Figure 2E, F). Given the uncertainties of absolute time, how local variability in $\delta^{13}C$ values in shallow-water environments can be expressed in stratigraphic records on local, regional, and global scales is vitally important for $\delta^{13}C$ chemostratigraphy (Figure 3).

3.1 Local Variability in Carbon Sources and Sinks

3.1.1 Riverine Input

One of the main sources of DIC to shallow-water environments is the discharge of rivers and groundwater to coastal settings (Figure 3). Contrary to the assumptions made above concerning a single isotopic value for riverine input (F_{riv}; Equations 5 and 6), there is a large range in the concentration and isotopic composition of DIC from rivers (even if the average value is well constrained). As one example, streams in Sweden yielded $\delta^{13}C_{DIC}$ values ranging between −28 to 0‰ (Campeau et al. 2017). This range is a product of upstream soil respiration, local mixtures of carbonate and organic carbon weathering, and variable degrees of stream water DIC equilibration with atmospheric CO_2. In general, rivers with low $\delta^{13}C_{DIC}$ values are often associated with siliciclastic-dominated catchments with high degrees of organic carbon respiration (Brunet et al. 2009), while higher $\delta^{13}C_{DIC}$ values often are associated with carbonate-dominated catchments (Barth et al. 2003; Das et al. 2005; Khadka et al. 2014).

The difference in $\delta^{13}C_{DIC}$ values between siliciclastic- and carbonate-dominated rivers may be important to consider when interpreting the stratigraphic trends in $\delta^{13}C_{carb}$ values in mixed carbonate-siliciclastic depositional systems (e.g., Blanco et al. 2020). On carbonate platforms adjacent to continental blocks, deposition is influenced by clastic sediment and nutrient influx from river drainage systems that migrate over time. The chemostratigraphic implications of such a depositional setting is that $\delta^{13}C$ values of carbonate that precipitates in coastal waters may vary as a function of local siliciclastic input. It may be possible to disentangle this effect by linking sedimentological facies observations with $\delta^{13}C_{carb}$ values (Blanco et al. 2020). However, while the influx of siliciclastic material largely is controlled by local fluvial-deltaic processes, it is possible that global climate and/or eustatic sea-level change can generate widespread changes in relative weathering of carbonate versus siliciclastic material (Harper et al. 2015). For example, increased weathering and diagenesis of carbonate platforms during glacial sea-level lowstands would likely increase the $\delta^{13}C$ value of global riverine inputs (Kump et al. 1999; Dyer et al. 2015). Alternatively, increased weathering and remineralization of organic-rich sediments may increase the input flux of ^{13}C depleted carbon and decrease the surface ocean $\delta^{13}C_{DIC}$ value (Figure 2C, D). While possibly global in scale, the impact of changes in riverine weathering fluxes is easily amplified in shallow-water platforms and epeiric seas (e.g., Holmden et al. 1998).

3.1.2 Submarine Groundwater Discharge

In addition to riverine input, submarine groundwater discharge also has measurable effects on surface water $\delta^{13}C_{DIC}$ values in coastal areas (Figure 3). In restricted environments such as Florida Bay and the Little Bahama Banks, submarine groundwater discharge leads to surface water $\delta^{13}C_{DIC}$ values down to −7‰ (Patterson and Walter 1994). These low values are a product of organic carbon respiration in the freshwater aquifer that flows towards the sea from platform interiors. In some coastal areas, such as in mangrove-dominated tidal creeks and reefs, the flux of groundwater discharge varies with the tides, with the highest inputs of groundwater at low tide, which results in high creek DIC concentrations (>3 mmol/kg; Maher et al. 2013). Surface water $\delta^{13}C_{DIC}$ values vary by up to 10‰ across a tidal cycle, from -8‰ at low tide to +2‰ at high tide (Maher et al. 2013). However, it is unlikely that the entire range in $\delta^{13}C_{DIC}$ is captured by carbonate precipitation, as the carbonate saturation state ($\Omega = (Ca^{2+} * CO_3^{2-})/K_{sp}$, where K_{sp} is the solubility product for aragonite) is also lowered with the input of respired and dissolved CO_2, thus predicting less precipitation during periods of high submarine groundwater discharge and low $\delta^{13}C_{DIC}$.

In spite of the potential lack of carbonate mineral precipitation at times of very low $\delta^{13}C_{DIC}$, stratigraphic records from carbonate platforms may still preserve $\delta^{13}C_{carb}$ signals that are modified by local submarine groundwater discharge. For example, if local discharge into coastal environments carries significant amounts of dissolved carbonate in solution, positive $\delta^{13}C$ "excursions" can be created in carbonate precipitating from these waters (Holmden et al. 2012).

3.1.3 Diurnal Productivity

Dissolved inorganic carbon in restricted shallow-marine environments can exhibit large diurnal fluctuations in $\delta^{13}C_{DIC}$ values, driven by daytime photosynthesis and nighttime respiration of organic matter (Figure 3). For example, surface waters in reefs on O'ahu show a range in $\delta^{13}C_{DIC}$ values of up to 5‰ across a diurnal cycle (Richardson et al. 2017). In such settings, carbonate precipitation occurs predominantly during the most productive parts of the diurnal cycle, recording $\delta^{13}C_{carb}$ values up to +7‰ in platform aragonite (Geyman and Maloof 2019). This pattern is due to the respiration of organic carbon at nighttime, lowering Ω and hindering carbonate precipitation. As a result, stratigraphic records from shallow-water environments may be biased towards values of $\delta^{13}C_{carb}$ that are higher than average surface water $\delta^{13}C_{DIC}$. If shallow platforms contribute significantly to the global carbon budget then, as the relative area of shallow shelves increases, the burial of ^{13}C-enriched carbonate will lower the average seawater $\delta^{13}C_{DIC}$ to satisfy global mass balance (Geyman and Maloof 2019).

3.1.4 Air–Sea Gas Exchange

Highly variable $\delta^{13}C_{DIC}$ values of shallow-water environments are possible due to slow rates of air–sea gas exchange, meaning that surface water $\delta^{13}C_{DIC}$ can diverge from equilibrium values with the overlying atmosphere (Lynch-Stieglitz et al. 1995). The exchange of CO_2 across the air–sea interface is also associated with kinetic isotope effects, with the preferential dissolution and degassing of ^{12}C (e.g., Wanninkhof 1985). Surface waters that have low DIC concentrations with net CO_2 invasion may therefore have lower $\delta^{13}C_{DIC}$ values compared to surface waters with high DIC concentration and net CO_2 degassing (Wanninkhof 1985). The isotope effects of air–sea gas exchange can dampen the $\delta^{13}C_{DIC}$ enrichment from the biological pump in surface waters, but in the modern ocean these kinetic effects are relatively small, leading to spatial variability in surface waters $\delta^{13}C_{DIC}$ of the open ocean of up to ~2‰ (Lynch-Stieglitz et al. 1995).

In contrast to the open ocean, kinetic isotope effects associated with CO_2 air–sea gas exchange can be pronounced in hypersaline and restricted environments (e.g., Clark et al. 1992; Lazar and Erez 1992; Beeler et al. 2020). In these settings, carbonate with $\delta^{13}C$ values between −25‰ and +15‰ have been documented (Figure 3), with rapid invasion of CO_2 leading to low $\delta^{13}C_{carb}$ values (Clark et al. 1992) and degassing of CO_2 leading to high $\delta^{13}C_{carb}$ values (Beeler et al. 2020). It is unclear if kinetic effects are expressed in stratigraphic $\delta^{13}C_{carb}$ records, but it has been suggested that these processes may have been more pronounced in

Precambrian platform environments, where abiotic or microbially mediated carbonate precipitation likely was more important (Ahm et al. 2019; Husson et al. 2020; Ahm et al. 2021).

3.1.5 Redox and Authigenic Mineralization

In both shallow and deep marine settings, $\delta^{13}C_{carb}$ values may be influenced by anoxic remineralization of organic carbon and consequent precipitation of in situ (authigenic) carbonate in the sediment pore space (Figure 3). More specifically, methanogenesis (methane production) and sulfate reduction within the sediment pile may lead to the precipitation of authigenic carbonate with extreme $\delta^{13}C_{carb}$ values. Methane (CH_4) has very low $\delta^{13}C$ values (~−60 to −80‰; Claypool and Kaplan 1974), meaning that its production leaves residual porewater DIC with heavy $\delta^{13}C$ values. In contrast to oxic respiration of organic matter, which lowers porewater Ω values and promotes carbonate dissolution, anoxic respiration (such as sulfate reduction) tends to increase Ω and rates of carbonate precipitation (Claypool and Kaplan 1974). For example, methane oxidation by sulfate reduction has been shown to produce carbonate cements with $\delta^{13}C_{carb}$ values down to <-50‰ (Hovland et al. 1987; Bohrmann et al. 1998; Naehr et al. 2000). In contrast, $\delta^{13}C_{carb}$ values up to +16‰ have been documented in stromatolitic and microbial carbonate as the result of methanogenesis (Birgel et al. 2015). Consequently, if representing a significant fraction of the bulk sediment, authigenic carbonates can produce local stratigraphic trends in $\delta^{13}C_{carb}$ values. In addition, if authigenic carbonate formation is a significant carbonate sink, it may have the potential to influence global $\delta^{13}C_{DIC}$ values (Bjerrum and Canfield 2004; Schrag et al. 2013; Barnes et al. 2020; Canfield et al. 2020).

The local processes discussed above (Section 3.1) are all associated with specific isotopic signals that may be expressed in the stratigraphic record of $\delta^{13}C_{carb}$ values (e.g., Pancost et al. 1999; Panchuk et al. 2006). Importantly, each of these processes will be amplified in restricted shallow-marine basins, and stratigraphic trends in $\delta^{13}C_{carb}$ and $\delta^{13}C_{org}$ values may originate from expansion and contraction of geochemically distinct surface waters, rather than global-scale changes in the carbon cycle (Holmden et al. 1998).

3.2 Carbonate Mineralogy, Fractionation, and Mixing

The $\delta^{13}C$ value of carbonate is fractionated relative to local surface water DIC, with the preferential uptake of ^{13}C into the carbonate mineral lattice (contrary to the assumption in Equation 5 that $\delta_{DIC} = \delta_{carb}$). The magnitude of fractionation varies for different carbonate polymorphs. Aragonite is more enriched in

^{13}C (~3‰) compared to calcite (~1‰; Romanek et al. 1992). As a result, stratigraphic changes in $\delta^{13}C_{carb}$ values can be produced by the mixing of aragonite, which is predominantly made in shallow-water carbonate factories like the Bahamas (Lowenstam and Epstein 1957), and calcite, the dominant polymorph of pelagic calcifiers (Bown et al. 2004). For example, it is possible to generate systematic changes in the fraction of platform versus pelagic-derived carbonate during periods of sea-level change, which may lead to coherent and reproducible stratigraphic changes in $\delta^{13}C$ values that are decoupled from changes to the global carbon cycle (Case Study 4.2; Swart and Eberli 2005; Swart 2008). Mixing of calcite and aragonite end-members, even if both formed from the same DIC pool, can also lead to "noise" in the stratigraphic $\delta^{13}C_{carb}$ record (e.g., Figure 2E,F).

In addition to mineralogical differences in fractionation factors, the $\delta^{13}C$ values of biogenic carbonates can also be affected by vital effects. Vital effects are related to both the local seawater chemistry and the metabolism of the living organism and influence the incorporation of carbon isotopes during biomineralization (e.g., Zeebe et al. 1999). Bulk carbonate $\delta^{13}C$ records can therefore be complicated by mixing biogenic materials with different fractionation factors and vital effects. For carbon isotope chemostratigraphy, these issues can in part be addressed though building records from single species – commonly done, for example, in foraminifera research.

Similarly to biogenic carbonate, there is considerable variability in isotopic values of organic matter and, hence, local differences in Δ_B (Equations 5 and 6; Hayes et al. 1999; Freeman 2001). The range of isotopic values of organic matter is related to organismal growth rates and the specific carbon fixation pathways employed (Freeman 2001). As a result, fractionations can be highly variable (even within a single species), and terrigenous, coastal, and marine organic carbon can be locally offset by >20‰ (Pancost et al. 1999; Freeman 2001; Oehlert et al. 2012). The net fractionation factor between local surface water DIC and organic carbon (Δ_B) is also dependent on the concentration of CO_2 in the ambient environment, with higher CO_2 levels leading to a more negative fractionation (reflecting preferential uptake of ^{12}C; Hollander and McKenzie 1991; Popp et al. 1998). Combined, these isotope effects are relevant to consider in shallow-water coastal environments where both DIC concentrations and DIC $\delta^{13}C$ values (and, hence, ambient CO_2) vary on semi-diurnal and diurnal timescales (Section 3.1.3).

Differences in the isotopic fractionation of organic matter can be observed in chemostratigraphic records, since the bulk isotopic composition of organic carbon at a given locality can be mixtures from different sources (Figure 3). In coastal environments especially, these inputs can be highly variable both in space and time, with transport and deposition of sediments from coastal environments into deeper waters causing stratigraphic changes in $\delta^{13}C$ values of both carbonate and organic carbon due to mixing (Prahl et al. 1994; Pancost et al. 1999; Oehlert and Swart 2014). Moreover, $\delta^{13}C_{org}$ values are also affected by both syn- and post-depositional processes where the preferential degradation of labile organic compounds may lead to an increase in the $\delta^{13}C$ values of residual organic carbon of up to 4–5‰ (Oehlert and Swart 2014).

3.3 Carbonate Diagenesis

As carbonate sediments are buried and lithified, they pass through several stages of diagenesis (e.g., meteoric, marine, burial), where primary metastable carbonate minerals dissolve and diagenetic minerals precipitate. The most common approach for predicting whether the isotopic value of a given element (e.g., C, O, Ca, trace metals) will be reset during lithification is to consider the comparative abundances of that element in altering fluids and in the sediment (Gross 1964; Allan and Matthews 1982; Banner and Hanson 1990). If an element has low abundance in the fluid relative to the sediment (such as for carbon in seawater compared to sediment), the sediment is more likely to retain its original isotopic composition (sediment-buffered diagenesis). The concentration of elements, such as carbon, in the pore fluid varies significantly across different diagenetic settings and stages. For example, in comparison to seawater (marine diagenesis) meteoric fluids tend to have much higher concentrations and lower $\delta^{13}C$ values of DIC due to the respiration of organic carbon within freshwater aquifers (freshwater phreatic zone; Swart 2015).

In general, in settings with high fluid flow rates (advection dominated), the cumulative fluid-to-rock ratio for carbon becomes high and the isotopic composition of the primary carbonate sediment can be reset (fluid-buffered diagenesis). Fluid-buffered diagenetic regimes are common in shallow-water and periplatform environments where flow rates are high and largely driven by buoyancy and geothermal convection (Figure 3, ~10 cm/yr; Kohout 1965; Henderson et al. 1999). In contrast, sediment-buffered diagenesis occurs in settings where fluid flow rates are low (diffusion dominated) or in settings with low fluid carbon concentrations. During sediment-buffered diagenesis, the isotopic composition of the pore fluid is in equilibrium with the sediment and does not have the potential to alter $\delta^{13}C_{carb}$ values. Sediment-buffered diagenesis is characteristic of the deep ocean seafloor where subsurface fluid flow is diffusion dominated (e.g., Fantle et al. 2010) or during late-stage burial diagenesis where porefluid tends to have reacted extensively with the host strata. As a result, if $\delta^{13}C$ values of carbonate rock are reset, this process likely occurs relatively early, within the first 100s meter below the seafloor, at burial temperatures <40°C (Staudigel and Swart 2019; Murray et al. 2021).

3.3.1 Calcium Isotopes

Other geochemical proxies can be measured along with $\delta^{13}C_{carb}$ to evaluate the degree of diagenetic alteration (e.g., carbonate $\delta^{18}O$ and Mn/Sr measurements). More recently, calcium isotopes ($\delta^{44/40}Ca$) have emerged as a powerful tool for disentangling the degree of fluid versus sediment-buffered diagenesis of $\delta^{13}C_{carb}$ values. The advantage of carbonate $\delta^{44/40}Ca$, in comparison to $\delta^{18}O$ values for example, is that the ratio of calcium in carbonate minerals relative to seawater is similar to that of carbon, which means that the two isotopic systems respond to diagenesis at similar fluid-to-rock ratios (Fantle and Higgins 2014; Ahm et al. 2018). Combining $\delta^{13}C_{carb}$ values with both $\delta^{44/40}Ca$ and Sr/Ca ratios can fingerprint different diagenetic end members (Figure 4C–D). This tool is useful because Ca isotope fractionation and Sr partitioning is sensitive to both carbonate mineralogy and precipitation rate (Gussone et al. 2005; Tang et al. 2008). Primary aragonite is more depleted in ^{44}Ca and enriched in Sr (with values of −1.5‰ and 10 mmol/mol, respectively) relative to primary calcite (−1‰ and ~1 mmol/mol). Diagenetic calcite or dolomite is characterized by lower Sr contents (<1 mmol/mol) and less fractionated $\delta^{44/40}Ca$ values, approaching ~0‰ at equilibrium with the pore fluids (Fantle and DePaolo 2007; Jacobson and Holmden 2008). As a result, sediment-buffered diagenesis will be labeled by low $\delta^{44/40}Ca$ values and relatively high Sr/Ca ratios, while fluid-buffered diagenesis will be labeled by high $\delta^{44/40}Ca$ values and low Sr/Ca ratios (Figure 4C–D; Higgins et al. 2018).

In addition to helping identify primary and diagenetic end members, $\delta^{44/40}Ca$ values can also shed light on the degree to which geographically disparate carbonate successions, which are often correlated using carbon isotope chemostratigraphy (especially in the Precambrian), reflect the globally averaged carbonate sink (Blättler and Higgins 2017). The main sink from the ocean is the burial of carbonate. Thus, a prediction for the global calcium cycle is that when in a steady state, the globally averaged calcium isotope composition of carbonate sediments should equal that of bulk silicate Earth (~ −1‰ on timescales >10^6 years; Skulan et al. 1997; Blättler and Higgins 2017). In other words, if the Ca cycle is in steady state, the $\delta^{44/40}Ca$ of the average carbonate sink has a predictable value (~ −1‰). Calcium isotopes can therefore be used not only to understand diagenesis, but when averaged across the globe, may also be used to evaluate if correlated stratigraphic sections reflect the average carbonate burial sink. For example, in scenarios where carbonate strata have both negative $\delta^{13}C$ values and $\delta^{44/40}Ca$

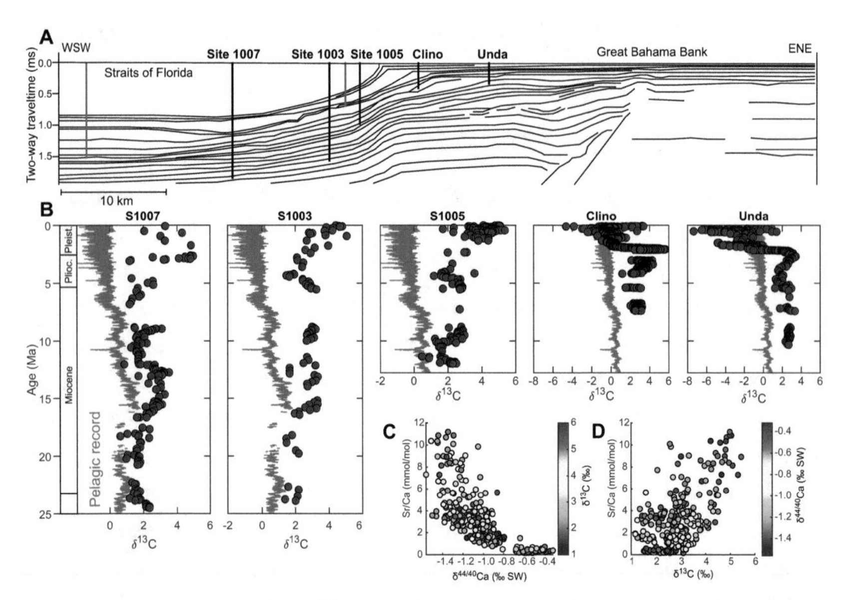

Figure 4 The variability in $\delta^{13}C_{carb}$ values from the Bahamas Transect (A), in cores taken across the bank-top and peri-platform slope, differ from the trends observed in deep-sea pelagic records (grey line; Westerhold et al. 2020). (B) Across the Bahamas transect (Eberli et al. 1997; Ginsburg 2001), the downcore trends are products of early marine diagenesis, meteoric alteration, and mixing between platform aragonite ($\delta^{13}C_{carb}$ ~6‰) and pelagic carbonate $\delta^{13}C_{carb}$ ~1‰). Notably, in peri-platform cores (S1007, S1005, S1003) the increasing trends in $\delta^{13}C_{carb}$ across the Plio-Pleistocene are also observed in platforms across the world (Swart 2008), and record mixing between platform aragonite, transported from the platform top, and both pelagic and/or diagenetic calcite and dolomite (Swart and Eberli, 2005; Higgins et al. 2018). The top of the cores from the bank top (Clino and Unda), record depleted $\delta^{13}C_{carb}$ values (down to −7‰), as a result of meteoric alteration during Plio-Pleistocene glacio-eustatic sea-level fall (Melim et al. 1995). (C) Downcore diagenetic recrystallization of aragonite can be tracked using $\delta^{44/40}Ca$ values (modern seawater as the reference standard) and Sr/Ca ratios. Primary platform aragonite sediments have low $\delta^{44/40}Ca$ values and high Sr/Ca ratios (D, high $\delta^{13}C_{carb}$), while diagenetic low-Mg calcite or dolomite have high $\delta^{44/40}Ca$ values and low Sr/Ca ratios (Higgins et al. 2018)

values more negative than bulk silicate Earth observed across several stratigraphic sections, mass balance requires that other carbonate sinks must exist that record more positive $\delta^{44/40}Ca$ values (e.g., authigenic cements, hydrothermal veins, or dolomitized carbonate platforms; Gussone et al. 2020). While it is not necessary that the $\delta^{13}C$ values of these unmeasured carbonates be different, it becomes less certain that the globally average carbonate sink does have a negative $\delta^{13}C$ value. The alternative scenario, where several stratigraphic sections with negative $\delta^{13}C$ values also coincide with an average $\delta^{44/40}Ca$ value of ~ -1‰, strengthens the argument for interpreting carbon isotope values in terms of global carbon fluxes (e.g., Figure 2A–D).

On shorter timescales, within the residence time of calcium in the ocean (<10^6 years), it is possible to have excursions in $\delta^{44/40}Ca$ values that reflect transient imbalances in calcium inputs and outputs. In this scenario, it is possible to have $\delta^{13}C$ and $\delta^{44/40}Ca$ excursions that correlate across the globe. Numerical models that include the coupled carbon cycle have demonstrated that the maximum transient $\delta^{44/40}Ca$ excursion that can occur from the combined effects of increased weathering (which can increase global ocean Ca concentration) and ocean acidification (which can depress burial) is ~−0.3‰ (Komar and Zeebe 2016). There are two implications for carbon isotope chemostratigraphy. First, if a $\delta^{13}C$ excursion correlates with a $\delta^{44/40}Ca$ excursion, then the duration of these excursions must be <10^6 years (if Ca residence time is similar to the modern). Second, the magnitude of change in $\delta^{44/40}Ca$ values has to be <0.3‰. If either is not true, then the excursions are difficult to interpret in terms of the global C and Ca cycles, indicating that local controls are most important for $\delta^{44/40}Ca$ and $\delta^{13}C$ values (Figure 3).

In Subsections 4.2–4.3, we highlight case studies that demonstrate the potential to disentangle global and local processes using a combination of carbonate $\delta^{44/40}Ca$ and $\delta^{13}C$ values.

4 Case Studies

4.1 Paleocene–Eocene Thermal Maximum

Perhaps the best example of a globally synchronous excursion in $\delta^{13}C$ values that can be tracked across shallow- to deep-water depositional settings is the Paleocene-Eocene Thermal Maximum (PETM, ~56 Ma). The PETM is associated with a negative carbon isotope excursion (~−3‰, recorded in both carbonate and organic carbon), and is interpreted as a product of rapid injection of ^{13}C-depleted carbon into the ocean-atmosphere system (~14,900 Pg C; Zachos et al. 2003, 2005; Haynes and Hönisch 2020), resulting in rapid greenhouse warming. Aided by high-resolution radiometric age constraints, biostratigraphy, and correlation by astronomical tuning, the PETM has been shown to be synchronous across deep-water, shallow-marine, and terrestrial environments, with $\delta^{13}C$ values returning to background over ~170 kyrs following the excursion (Röhl et al. 2007; Westerhold et al. 2018; Zeebe and Lourens 2019).

Furthermore, the PETM is also associated with ocean deoxygenation (Zhou et al. 2014; Clarkson et al. 2021), global sea-level change (Sluijs et al. 2008), and widespread dissolution of pelagic carbonates (Zachos et al. 2005), with carbonate deposition often replaced by deposition of clay-rich ooze in the minimum of the excursion.

Despite the well-constrained global extent of the PETM, the amount and source of carbon released across the excursion is debated due to large differences in the magnitude of the $\delta^{13}C$ change recorded across different depositional environments (McInerney and Wing 2011). For example, the magnitude of the $\delta^{13}C$ excursion in terrestrial environments and shallow-water platforms is larger and more variable than in the pelagic records (between −15 to −2‰, respectively, Bowen et al. 2015; Li et al. 2020, 2021). The large magnitude of the PETM excursion in carbonate platform environments is likely a consequence of local effects that were amplified by relative sea-level change and increased platform restriction (Subsection 3.1). Moreover, in shallow-water settings, the excursion is often associated with a facies change from carbonate to mixed-siliciclastic dominated deposition (McInerney and Wing 2011; Li et al. 2021), indicating that hydrological changes, increased riverine input of siliciclastic material, and remineralized organic carbon, could have attributed to the locally lower $\delta^{13}C_{carb}$ values (Subsection 3.1).

In addition to local effects in shallow-water and terrestrial environments, the magnitude of the PETM also varies considerably in pelagic carbonate sediments deposited in the open ocean. For example, mixed-layer foraminifera record a larger and more rapid $\delta^{13}C$ excursion (3.5–4‰) compared to benthic and deep-water dwelling species (~2‰, Zachos et al. 2007). These differences may reflect the spatial response of the ocean to a rapid carbon injection, with the surface ocean responding more rapidly than the deep, potentially indicating that the release of carbon was faster than the timescale for vertical mixing of water masses (Zachos et al. 2007). However, carbonate biomineralization is associated with vital effects that can result in significant differences between $\delta^{13}C_{carb}$ values of individual pelagic species (Subsection 3.2). As a result, even well-constrained pelagic carbonate records are associated with significant isotopic variability that requires careful evaluation.

While the PETM is arguably the best example of when carbon isotope chemostratigraphy works, this globally recorded event also readily demonstrates the problems with trying to infer changes in global mass fluxes (f_{org}; Subsection 2.2) directly from $\delta^{13}C$ values, owing to the variability observed in both organic and carbonate records (especially from shallow water depositional environments). Thus, lessons from the PETM are particular important to consider in studies of ancient platform carbonates – specifically, during time periods where equivalent deep sea pelagic carbonate records are unavailable.

4.2 The Great Bahamas Bank

The Bahamas carbonate platform is one of the best studied examples of the effects of local carbon cycling, restriction, and carbonate diagenesis on $\delta^{13}C$ values. The Bahamas Drilling Project and the Ocean Drilling Program (ODP Leg 166), acquired a transect of cores across the Great Bahamas Bank west of Andros Island, including the platform top (Clino and Unda), adjacent slope (peri-platform), and the basinal environments in the deep waters of the Strait of Florida (Figure 4A; Eberli et al. 1997; Ginsburg 2001). Chemostratigraphies measured from these cores provide important constraints on the geochemical signatures of platform progradation, oscillating sea level, and carbonate diagenesis, while largely being disconnected from global $\delta^{13}C_{DIC}$ values.

Carbonate sediments from the Great Bahama Bank and peri-platform consists of a mixture sourced from the bank top (platform aragonite) and open ocean (pelagic calcifiers; Eberli et al. 1997; Ginsburg 2001). Stratigraphic variations in the proportion of platform and pelagic carbonate have been linked to changes in eustatic sea level, with pelagic carbonate dominating during periods of low sea level and platform-derived carbonate dominating during high sea level (Swart and Eberli 2005). During sea-level lowstands, carbonate platform tops would be exposed, thus shutting down aragonite production and export; conversely, during highstands, the platform is flooded and a highly productive source of aragonite mud (Schlager et al. 1994). The significance of sediment mixing, with respect to carbon isotope chemostratigraphy, is demonstrated by a pronounced correlation between carbonate $\delta^{13}C$ values and the percentage of aragonite in cores across the Bahamas transect (Subsection 3.2; Swart and Eberli 2005). Aragonite mud, produced by calcareous green algae such as *Halimeda* on the shallow platform top, has $\delta^{13}C$ values of ~+6%, while pelagic calcite from coccolith and foraminifera tests have values ~+1‰ (Lowenstam and Epstein 1957; Swart and Eberli 2005). The elevated $\delta^{13}C$ values of platform aragonite mud are a product of the larger fractionation factor for aragonite compared to calcite, in addition to intense diurnal productivity that elevates the surface water $\delta^{13}C_{DIC}$ and carbonate saturation in the day time when aragonite precipitates (Swart and Eberli 2005; Geyman and Maloof 2019). As a consequence, production and shedding of aragonite from the platform during highstands (Schlager et al. 1994) is recorded by a positive excursion in carbonate $\delta^{13}C$ values in peri-platform and basinal sediments

(Figure 4). The influence of highstand aragonite shedding on downslope $\delta^{13}C$ values is a global phenomenon, owing to the of high-amplitude glacioeustatic sea-level changes in the Plio-Pleistocene (Figure 4; Swart and Eberli 2005; Swart 2008). In contrast, carbonate sediments from the platform top (cores Clino and Unda) also record high aragonite $\delta^{13}C$ values, but during the Plio-Pleistocene these sediments were altered by meteoric fluids due to sea-level fall and platform exposure (seelater in this Element). As a result, positive $\delta^{13}C$ values are only recorded in the lower part of the platform cores (Figure 4). These positive "excursions" are important examples of carbonate $\delta^{13}C$ values not reflecting changes in global carbon fluxes, in spite of being broadly correlative across the globe (Swart 2008).

Attributing excursions in $\delta^{13}C$ values to mixing of aragonite and pelagic calcite is complicated by effects from early marine diagenesis, but may be disentangled using measurements of $\delta^{44/40}Ca$ values (Subsection 3.3; Higgins et al. 2018). Geothermal temperature gradients drive the advection of seawater into the platform interior from the slope (Kohout 1965; Henderson et al. 1999), resulting in significant fluid-buffered diagenesis of periplatform sediments (Melim et al. 2002; Higgins et al. 2018). These advected fluids initially have $\delta^{13}C$ and $\delta^{44/40}Ca$ values that reflect the open ocean (+1 and 0‰, respectively). As a result, fluid-buffered diagenesis of platform-derived aragonite results in the resetting of carbon $\delta^{13}C$ values from +5 to +1‰ (Figure 4D; Ahm et al. 2018; Higgins et al. 2018). The process is recorded in the slope of the Great Bahamas Bank (Sites 1003 and 1007) as a downcore negative trend in $\delta^{13}C$ values, caused by the progressive dissolution of metastable platform aragonite and precipitation of diagenetically stable low magnesium calcite or dolomite. Progressive aragonite replacement is also tracked by carbonate $\delta^{44/40}Ca$ values and Sr/Ca ratios that correlate with carbonate $\delta^{13}C$ values in the upper ~150 m of the peri-platform cores (Figure 4C). Primary platform aragonite has low $\delta^{44/40}Ca$ values (~−1.5‰) and high Sr/Ca ratios (~10 mmol/mol) due to the higher partition coefficients of Sr and ^{40}Ca for aragonite than calcite (Tang et al. 2008). During aragonite replacement, $\delta^{44/40}Ca$ is reset to higher values (~−0.5‰) and Sr/Ca ratios decrease (<1 mmol/mol; Higgins et al. 2018).

In addition to invigorating the advection of marine fluids into the platform interior, glacio-eustatic sea-level changes in the Plio-Pleistocene resulted in extended periods of exposure of the platform top (Vahrenkamp et al. 1991; Melim et al. 1995). During glacial maxima, platforms are exposed and the freshwater aquifers within the exposed islands expand, causing meteoric diagenesis of previously deposited aragonite mud (Figure 4; cores Clino and Unda). In the meteoric lens, groundwater acquires carbon from the degradation of organic matter in the surrounding sediment, which leads to the release of isotopically light

CO_2 and the promotion of aragonite dissolution (Allan and Matthews 1977). Due to the influence of respired organic carbon in the freshwater lens, $\delta^{13}C$ values of re-precipitated low magnesium calcite or dolomite can have very low values, consistent with observations from the top of the Clino and Unda cores (Figure 4B). During Pleistocene glacial-interglacial transitions, the Bahamas Bank has been exposed repeatedly, leading to a deep profile (+100 m) meteorically altered sediments in platform top carbonates (Melim et al. 1995; Swart and Eberli 2005). With regard to carbon isotope chemostratigraphy, widespread synchronous negative excursions in carbonate $\delta^{13}C$ values are expected as a result of sea-level fall and increase meteoric diagenesis of platform tops (provided significant organic carbon respiration occurs in the freshwater lens). These excursions can be useful stratigraphic markers for correlation, specifically during “ice-house periods” characterized by significant sea-level change, but the $\delta^{13}C_{carb}$ would not be tracking global $\delta^{13}C_{DIC}$ (Allan and Matthews 1977; Dyer et al. 2015, 2017). The most robust methods to fingerprint negative excursions as meteoric is by comparing other geochemical signatures (e.g., negative $\delta^{13}C$ correlating with negative $\delta^{18}O$, or high Mn concentrations), in addition to petrographic observations of meteoric cements and sedimentological features such as exposure surfaces and root casts (Melim et al. 2001; Swart 2015; Oehlert and Swart 2019).

Owing to the extensive work on carbonate geochemistry from the Bahamas (e.g., Ginsburg 2001; Swart and Eberli 2005; Swart 2008; Oehlert et al. 2012; Oehlert and Swart 2014; Higgins et al. 2018), it is well established that Bahamian carbonate $\delta^{13}C$ values do not reflect the $\delta^{13}C_{DIC}$ values of a well-mixed ocean in equilibrium with the atmosphere. While carbon isotope chemostratigraphy of pelagic deep-sea sediments may better reflect the global carbon cycle (e.g., Westerhold et al. 2020), the application of chemostratigraphy to ancient platform carbonates is not straightforward.

4.3 The Neoproterozoic

Due to the lack of index fossils useful for biostratigraphy, carbon isotope chemostratigraphy has been applied widely to correlate Neoproterozoic carbonate successions (1000–541 Ma; Knoll et al. 1986; Halverson et al. 2005). The variability in $\delta^{13}C$ values from Neoproterozoic carbonates dwarfs that of the Cenozoic deep sea record (population standard deviation = 4.7‰ vs. 0.6‰ respectively; Figure 5). The Neoproterozoic record is characterized by high baseline values of +5–10‰ that are interrupted by dramatic negative excursions of values down to −15‰ (Figure 5). The origin of the Neoproterozoic carbon isotope excursions is still widely debated, because $\delta^{13}C$ values below −5‰ cannot be explained by a traditional steady state carbon cycle model. Namely, in Equation 6, inserting values for δ_{carb} that are below values of δ_{riv} will result in a negative number for the fraction of organic carbon buried. As a consequence, several studies have suggested alternative models for generating such negative $\delta^{13}C$ values, including (1) short-term transient perturbations to the carbon cycle (e.g., Schrag et al. 2002; Rothman et al. 2003; Bjerrum and Canfield 2011), (2) carbonate diagenesis (Knauth and Kennedy 2009; Derry 2010), or (3) the formation of authigenic carbonate minerals (Tziperman et al. 2011; Schrag et al. 2013; Laakso and Schrag 2020). However, these models do not consider the possibility of broadly synchronous shifts in $\delta^{13}C_{DIC}$ values of *platform* waters (Ahm et al. 2019, 2021; Crockford et al. 2020), that may differ between individual basins on timescales $>10^5$ years, and therefore may not directly record changes in global DIC.

The largest negative carbon isotope excursion in the Neoproterozoic is the Ediacaran Shuram-Wonoka excursion. This excursion is observed across the globe (e.g., Oman, Australia, Death Valley, Northwest Canada, Siberia) with minima $\delta^{13}C_{carb}$ values of ~−15‰ (Figure 5). Recent Re-Os ages confirm that the excursion is broadly synchronous, ranging between ~574±4.7 to 567±3.0 Ma (Rooney et al. 2020). These geochronological constraints, however, do not necessarily require that the excursion represent a global carbon cycle perturbation (*sensu* Figure 2). Similarly to observations from recent platform carbonate, $\delta^{44/40}Ca$ data from the Wonoka Formation (South Australia) suggests that much of the carbonate that make up the anomaly was aragonite, originally formed in platformal settings, transported down slope, and recrystallized to low-Mg calcite during sediment-buffered diagenesis (Husson et al. 2015). Large systematic changes in $\delta^{44/40}Ca$ values (between −0.5 to −2.0%) occur across the excursion (Figure 2B,C), which is inconsistent with changes in the global calcium cycle and indicate that these sediments cannot represent the average carbonate burial sink (on timescale $>10^6$ yrs; Section 3.3; Blättler and Higgins 2017). Moreover, in-situ measurements of $\delta^{13}C$ values via secondary ion mass spectrometry on individual carbonate grains, representing both authigenic phases and transported platform sediments, yielded a large range (from +5 to −15‰) from a single hand sample (Husson et al. 2020). These results suggest that the negative carbonates recording the Shuram excursion are recording local carbon cycling in a specific Ediacaran surface environment and not changes in the $\delta^{13}C_{DIC}$ value of average seawater. The Shuram excursion may therefore be a broadly synchronous chemostratigraphic marker that track changes to global climate, tectonics, or sea level over millions of years (Rooney et al. 2020), without tracking global $\delta^{13}C_{DIC}$ (Busch et al., 2022).

Similar to the Shuram excursion, correlated changes in both carbonate $\delta^{13}C$ and $\delta^{44/40}Ca$ values are found in the globally distributed, basal Ediacaran carbonates (~635 Ma) that "cap" glacial deposits created during the pan-glacial

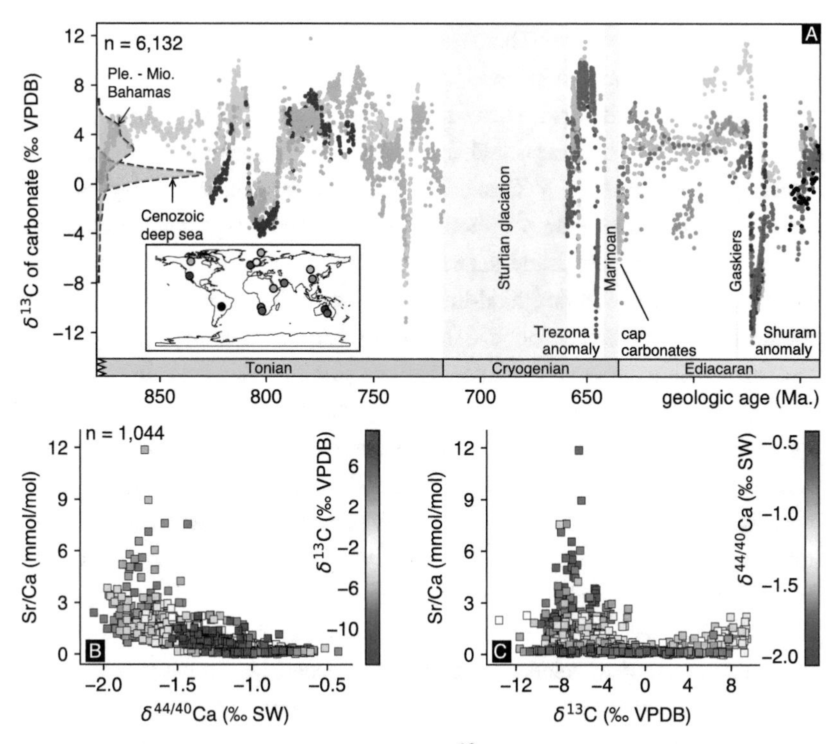

Figure 5 (A) A global compilation of δ^{13}C values from Neoproterozoic carbonates. The Tonian and Ediacaran age models are derived from Swanson-Hysell et al. (2015) and Rooney et al. (2020), with age constraints for the Cryogenian from Hoffman et al. (2017). Data points are color coded to approximate geographic locality shown in the inset map. Kernel density estimates of distributions of δ^{13}C values from both the Bahamas (red, Pleistocene-Miocene in age; Melim et al. 1995; Swart and Eberli, 2005) and deep sea sediment cores (blue, Cenozoic in age; Westerhold et al. 2020) are also shown. The height of each density estimate is arbitrary. (B, C) Cross-plots of $\delta^{44/40}$Ca, δ^{13}C and Sr/Ca values from carbonates with notable excursions in carbon isotopes: the Trezona and Shuram anomalies (Husson et al. 2015; Ahm et al. 2021), and basal Ediacaran cap carbonates (Ahm et al. 2019).

"Snowball Earth" climate state (Hoffman et al. 2017). In these cap carbonates, δ^{13}C values in limestones reach −6‰, coinciding with low $\delta^{44/40}$Ca values (−2%) and high Sr/Ca ratios (~4 mmol/mol), consistent with signatures of sediment-buffered recrystallization of platform aragonite (Ahm et al. 2019). In contrast, dolostone portions of the cap sequence have higher δ^{13}C and $\delta^{44/40}$Ca values, approaching values of modern seawater, suggesting that fluid-buffered dolomitization in reaction with Ediacaran seawater was responsible for resetting the primary low $\delta^{13}C$ values. These results indicate that the surface waters of platforms, where aragonite was forming, were significantly depleted in ^{13}C (−6‰), while the open ocean had $\delta^{13}C_{DIC}$ values close to 0‰. Multiple cap carbonate sections measured across individual basins reveal systematic spatial gradients in both $\delta^{13}C$ and $\delta^{44/40}Ca$ that are related to original basin geometry. The lowest values in both isotopic systems, representing sediment-buffered diagenesis, are found in sections from the platform interior, while the highest values, representing fluid-buffered diagenesis, are characteristic of the platform edge and upper slope (Ahm et al. 2019). These spatial trends are consistent with patterns of geothermal convection of fluids through recent platforms (Kohout 1965; Henderson et al. 1999), where warm, buoyantly rising fluid within the platform interior leads to compensatory flow of cold seawater into the sediment pile from the platform slope. An implication of these observations is that $\delta^{13}C$ values of dolostone, formed via fluid-buffered diagenesis with seawater, may be more reliable recorders of average seawater in comparison to shallow-water carbonate from the platform interior (Ahm et al. 2019; Hoffman and Lamothe 2019), although the fractionation factor associated with dolomitization and different carbonate mineralogies should be accounted for (Section 3.2).

The results from the cap carbonates adds to the growing body of evidence that Neoproterozoic shallow water environments were characterized by large fluctuations in $\delta^{13}C$ values, decoupled from the average ocean value. Various mechanisms have been proposed to explain the origin of very depleted $\delta^{13}C$ values in Neoproterozoic platform interior carbonates but, as of yet, no single hypothesis is accepted widely. Potentially important processes include nucleation kinetics associated with nonskeletal carbonate production (Hoffman and Lamothe 2019), the influence of microbial mats on the precipitation of carbonate, and the rapid invasion of CO_2 during periods of intense productivity (Subsection 3.1.4; Lazar and Erez 1992). Whether or not these mechanisms could create broadly synchronous $\delta^{13}C$ excursions in shallow-water carbonates globally, and what boundary conditions and forcings are necessary for such synchronous changes, is an important avenue of ongoing research in Earth history.

5 Future Prospects

The original framework of carbon isotope chemostratigraphy built on the assumption that $\delta^{13}C_{carb}$ values directly record the $\delta^{13}C_{DIC}$ values of average global seawater, in a well-mixed ocean in equilibrium with the atmosphere (e.g., Broecker 1970; Kump and Arthur 1999). Research though the last twenty years, however, has documented the importance of considering local variability in

$\delta^{13}C_{carb}$ in platform environments (e.g., Holmden et al. 1998; Swart and Eberli 2005; Oehlert and Swart 2014; Bowen et al. 2015; Li et al. 2020, 2021) These findings have demonstrated that carbon isotope chemostratigraphy from shallow-water strata rarely represents perfect records of changes in global $\delta^{13}C_{DIC}$ values or the relative burial flux of organic carbon, f_{org}.

It is likely that stratigraphic changes in $\delta^{13}C_{carb}$ broadly correlate both within basins and across continents, driven by external forcings such as global climate change, tectonics, and sea level. To a first order, carbon isotope chemostratigraphy has provided robust correlation tie points, with uncertainties from hundred thousand to millions of years (Rooney et al. 2020; Swanson-Hysell et al. 2015) – similarly to the time constraints provided by biostratigraphy applied in shallow-water environments (Holland 2020). Importantly, these uncertainties should be considered when making arguments related to the global carbon cycle, which requires time constraints on scales of $<10^5$ years in order to estimate average carbon burial fluxes.

Going forward, to help constrain globally averaged carbonate burial on timescales $>10^6$ years, measurements of carbonate $\delta^{44/40}Ca$ values from thick carbonate successions can be a powerful approach (Section 3.3; Blättler and Higgins 2017; Higgins et al. 2018). Modeling studies of carbonate diagenesis that combine measurements of $\delta^{13}C$, $\delta^{44/40}Ca$, and Sr/Ca ratios, have demonstrated that by embracing the diagenetic history of ancient carbonate it is possible to derived more accurate records of seawater chemistry (Ahm et al. 2018). Specifically, carbonate successions that have experienced early marine diagenesis and/or dolomitization may be important, and yet-to-be explored, archives of ancient seawater chemistry (Ahm et al. 2019; Hoffman and Lamothe 2019; Crockford et al. 2020).

By accepting that shallow-water sedimentary records are significantly modified by local carbon cycle dynamics, in addition to carbonate diagenesis, we may find that $\delta^{13}C$ measurements instead reflect the local fingerprints of important climatic and evolutionary processes expressed in shallow-water settings (e.g., Geyman and Maloof 2019). Better understanding local carbon cycling and syn-depositional processes in platform environments is of vital importance for the next generation of carbon isotope chemostratigraphers.

6 Designated Key Papers

The references that follow are a selected list (ordered by year of publication) of the "classics" and "best of the new" that specifically evaluate the implications of diagenesis and local controls on carbon isotope chemostratigraphy:

- Allan, J.R., Matthews, R.K., 1977. Carbon and oxygen isotopes as diagenetic and stratigraphic tools: Surface and subsurface data, Barbados, West Indies. Geology 5 (1), 16–20. *One of the first studies to show the impacts of diagenesis on carbon isotope chemostratigraphy.*
- Banner, J.L., Hanson, G.N., 1990. Calculation of simultaneous isotopic and trace element variations during water-rock interaction with applications to carbonate diagenesis. Geochimica et Cosmochomica Acta 54 (11), 3123–37. *Fundamental study that demonstrates the sensitivity of carbonate $\delta^{13}C$ values to alteration using a fluid-rock interaction model.*
- Patterson, W.P., Walter, L.M., 1994. Depletion of ^{13}C in seawater ΣCO_2 on modern carbonate platforms: Significance for the carbon isotopic record of carbonates. Geology 22, 885–8. *One of the first studies to show significant isotopic variation of modern bank top waters, due to the respiration of marine and terrestrial organic matter.*
- Holmden, C., Creaser, R.A., Muehlenbacks et al. 1998. Isotopic evidence for geochemical decoupling between ancient epeiric seas and bordering oceans: Implications for secular curves. Geology 26 (6), 567–70. *One of the first studies to discuss the implications of local variations in geochemical aquafacies on secular carbon isotope trends in ancient platforms.*
- Swart, P.K., Eberli, G., 2005. The nature of the $\delta^{13}C$ of periplatform sediments: Implications for stratigraphy and the global carbon cycle. Sedimentary Geology 175 (1–4), 115–29. *Study that contextualizes the findings of the Bahamas drilling project and ODP Leg 166, and discuss the implications for carbon isotope chemostratigraphy.*
- Oehlert, A.M. Swart, P.K., 2014. Interpreting carbonate and organic carbon isotope covariance in the sedimentary record. Nature Communications 5 (1), 4672. *This study showed that strong covariance between carbonate and organic $\delta^{13}C$ value can be produced by diagenetic mechanisms and introduction of terrestrial organic matter.*
- Higgins, J.A., Blättler, C.L., Lundstrom, E.A. et al, 2018. Mineralogy, early marine diagenesis, and the chemistry of shallow-water carbonate sediments. Geochimica et Cosmochimica Acta 220, 512–34. *This study demonstrates that early-marine diagenesis has large impact on geochemical records of $\delta^{44}Ca$ and $\delta^{13}C$ values from platform settings.*
- Geyman, E.C., Maloof, A.M., 2019. A diurnal carbon engine explains ^{13}C-enriched carbonates without increasing the global production of oxygen. PNAS 116 (49), 24433–24439. *This study demonstrates that highly ^{13}C-enriched carbonate can be produced in platform environments without increasing the local or global burial flux of organic carbon.*

- Hoffman, P.F., Lamothe, K.G., 2019. Seawater-buffered diagenesis, destruction of carbon isotope excursions, and the composition of DIC in Neoproterozoic oceans. PNAS 116 (38), 18874–79. *This study finds that large spatial gradients in $\delta^{13}C$ values exist across Neoproterozoic carbonate platforms as a result of early marine diagenesis.*

References

Ahm, Anne-Sofie C, Christian J Bjerrum, Clara L Blättler, Peter K Swart, and John A Higgins. 2018. "Quantifying Early Marine Diagenesis in Shallow-Water Carbonate Sediments." *Geochimica et Cosmochimica Acta* 236: 140–59.

Ahm, Anne-Sofie C, Christian J Bjerrum, and Paul F Hoffman et al. 2021. "The Ca and Mg Isotope Record of the Cryogenian Trezona Carbon Isotope Excursion." *Earth and Planetary Science Letters* 568 (August): 117002.

Ahm, Anne-Sofie C, Adam C Maloof, and Francis A Macdonald et al. 2019. "An Early Diagenetic Deglacial Origin for Basal Ediacaran 'Cap Dolostones'." *Earth and Planetary Science Letters* 506: 292–307.

Allan, JR, and Richard K Matthews. 1977. "Carbon and Oxygen Isotopes as Diagenetic and Stratigraphic Tools: Surface and Subsurface Data, Barbados, West Indies." *Geology* 5 (1): 16–20.

Allan, JR, and Richard K Matthews. 1982. "Isotope Signatures Associated with Early Meteoric Diagenesis." *Sedimentology* 29 (6): 797–817.

Banner, Jay L, and Gilbert N Hanson. 1990. "Calculation of Simultaneous Isotopic and Trace Element Variations During Water-Rock Interaction with Applications to Carbonate Diagenesis." *Geochimica et Cosmochimica Acta* 54 (11): 3123–37.

Barnes, Ben Davis, Jon M Husson, and Shanan E Peters. 2020. "Authigenic Carbonate Burial in the Late Devonian–Early Mississippian Bakken Formation (Williston Basin, USA)." *Sedimentology* 67 (4): 2065–94.

Barth, Johannes A. C., Cronin, Aidan A., Dunlop, J, and Kalin, Robert, M. 2003. "Influence of Carbonates on the Riverine Carbon Cycle in an Anthropogenically Dominated Catchment Basin: Evidence from Major Elements and Stable Carbon Isotopes in the Lagan River (N. Ireland)." *Chemical Geology* 200 (3): 203–16.

Beeler, Scott R, Fernando J Gomez, and Alexander S Bradley. 2020. "Controls of Extreme Isotopic Enrichment in Modern Microbialites and Associated Abiogenic Carbonates." *Geochimica et Cosmochimica Acta* 269: 136–49.

Berner, Robert A. 2006. "GEOCARBSULF: A Combined Model for Phanerozoic Atmospheric O_2 and CO_2." *Geochimica et Cosmochimica Acta* 70 (23): 5653–64.

Birgel, D, P Meister, R Lundberg et al. 2015. "Methanogenesis Produces Strong ^{13}C Enrichment in Stromatolites of Lagoa Salgada, Brazil: A Modern

Analogue for Palaeo-/Neoproterozoic Stromatolites?" *Geobiology* 13 (3): 245–66.

Bjerrum, Christian J, and Donald E Canfield. 2004. "New Insights into the Burial History of Organic Carbon on the Early Earth." *Geochemistry, Geophysics, Geosystems* 5 (8).

Bjerrum, Christian J, and Donald E. Canfield. 2011. "Towards a Quantitative Understanding of the Late Neoproterozoic Carbon Cycle." *Proceedings of the National Academy of Sciences* 108 (14): 5542.

Blättler, Clara L, and John A Higgins. 2017. "Testing Urey's Carbonate–Silicate Cycle Using the Calcium Isotopic Composition of Sedimentary Carbonates." *Earth and Planetary Science Letters* 479: 241–51.

Bohrmann, Gerhard, Jens Greinert, Erwin Suess, and Marta Torres. 1998. "Authigenic Carbonates from the Cascadia Subduction Zone and Their Relation to Gas Hydrate Stability." *Geology* 26 (7): 647–50.

Bowen, Gabriel J, Bianca J Maibauer, Mary J Kraus et al. 2015. "Two Massive, Rapid Releases of Carbon During the Onset of the Palaeocene–Eocene Thermal Maximum." *Nature Geoscience* 8 (1): 44–7.

Bown, Paul R, Jackie A Lees, and Jeremy R Young. 2004. "*Calcareous Nannoplankton Evolution and Diversity Through Time*." In Hans R Thierstein and Jeremy R Young, "Coccolithophores: From Molecular Processes to Global Impact" 481–508. Berlin, Heidelberg: Springer.

Broecker, Wallace S. 1970. "A Boundary Condition on the Evolution of Atmospheric Oxygen." *Journal of Geophysical Research* 75 (18): 3553–7.

Brunet, Frédéric, Dubois, Krystel, Veizer, Jan et al. 2009. "Terrestrial and Fluvial Carbon Fluxes in a Tropical Watershed: Nyong Basin, Cameroon." *Chemical Geology* 265 (3): 563–72.

Busch, James F., Hodgin, Eben B., Ahm, Anne-Sofie C., Husson, Jon M., Macdonald, Francis A., Bergmann, Kristin D., Higgins, John A., Strauss, Justin V. 2022. "Global and local drivers of the Ediacaran Shuram carbon isotope excursion." Earth and Planetary Science Letters 579 (February), 117368.

Campeau, Audrey, Marcus B Wallin, Reiner Giesler et al. 2017. "Multiple Sources and Sinks of Dissolved Inorganic Carbon Across Swedish Streams, Refocusing the Lens of Stable C Isotopes." *Scientific Reports* 7 (1): 1–14.

Canfield, Donald E, Andrew H Knoll, Simon W Poulton, Guy M Narbonne, and Gregory R Dunning. 2020. "Carbon Isotopes in Clastic Rocks and the Neoproterozoic Carbon Cycle." *American Journal of Science* 320 (2): 97–124.

Clark, Ian D, Jean-Charles Fontes, and Peter Fritz. 1992. "Stable Isotope Disequilibria in Travertine from High pH Waters: Laboratory Investigations and Field Observations from Oman." *Geochimica et Cosmochimica Acta* 56 (5): 2041–50.

Clarkson, Matthew O, Timothy M Lenton, Morten B Andersen et al. 2021. "Upper Limits on the Extent of Seafloor Anoxia During the PETM from Uranium Isotopes." *Nature Communications* 12 (1): 399.

Claypool, George E, and Isaac R Kaplan. 1974. "The Origin and Distribution of Methane in Marine Sediments." In *Natural Gases in Marine Sediments*, ed. Isaac R Kaplan, 99–139. Springer.

Craig, Harmon. 1953. "The Geochemistry of the Stable Carbon Isotopes." *Geochimica et Cosmochimica Acta* 3 (2–3): 53–92.

Crockford, Peter W, Marcus Kunzmann, Clara L Blättler et al. 2020. "Reconstructing Neoproterozoic Seawater Chemistry from Early Diagenetic Dolomite." *Geology*..

Das, Anirban, Krishnaswami, S, and Bhattacharya, Sourendra K. 2005. "Carbon Isotope Ratio of Dissolved Inorganic Carbon (DIC) in Rivers Draining the Deccan Traps, India: Sources of DIC and Their Magnitudes." *Earth and Planetary Science Letters* 236 (1): 419–29.

Derry, Louis A. 2010. "A Burial Diagenesis Origin for the Ediacaran Shuram–Wonoka Carbon Isotope Anomaly." *Earth and Planetary Science Letters* 294 (1–2): 152–62.

Dyer, Blake, John A Higgins, and Adam C Maloof. 2017. "A Probabilistic Analysis of Meteorically Altered δ^{13}c Chemostratigraphy from Late Paleozoic Ice Age Carbonate Platforms." *Geology* 45 (2): 135–8.

Dyer, Blake, Adam C Maloof, and John A Higgins. 2015. "Glacioeustasy, Meteoric Diagenesis, and the Carbon Cycle During the Middle Carboniferous." *Geochemistry, Geophysics, Geosystems* 16 (10): 3383–99.

Eberli, Gregor P, P K Swart, DF McNeill et al. 1997. "A Synopsis of the Bahamas Drilling Project: Results from Two Deep Core Borings Drilled on the Great Bahama Bank." In *Proceedings of the Ocean Drilling Program, Initial Reports*, 166: 23–41.

Fantle, Matthew S, and Donald J DePaolo. 2007. "Ca Isotopes in Carbonate Sediment and Pore Fluid from ODP Site 807A: The ca^2+(aq)–Calcite Equilibrium Fractionation Factor and Calcite Recrystallization Rates in Pleistocene Sediments." *Geochimica et Cosmochimica Acta* 71 (10): 2524–46.

Fantle, Matthew S, and John Higgins. 2014. "The Effects of Diagenesis and Dolomitization on Ca and Mg Isotopes in Marine Platform Carbonates: Implications for the Geochemical Cycles of Ca and Mg." *Geochimica et Cosmochimica Acta* 142: 458–81.

Fantle, Matthew S, K. M. Maher, and D. J. DePaolo. 2010. "Isotopic Approaches for Quantifying the Rates of Marine Burial Diagenesis." *Reviews of Geophysics* 48 (3).

Freeman, Katherine H. 2001. “Isotopic Biogeochemistry of Marine Organic Carbon.” *Reviews in Mineralogy and Geochemistry* 43 (1): 579–605.

Geyman, Emily C, and Adam C Maloof. 2019. “A Diurnal Carbon Engine Explains ^{13}C-Enriched Carbonates Without Increasing the Global Production of Oxygen.” *Proceedings of the National Academy of Sciences* 116 (49): 24433–9.

Ginsburg, Robert N., 2001. Subsurface Geology of a Prograding Carbonate Platform Margin, Great Bahama Bank: Results of the Bahamas Drilling Project. SEPM Society for Sedimentary Geology Volume 70.

Gregor, B. 1970. “Denudation of the Continents.” *Nature* 228 (5268): 273–5.

Gross, M Grant. 1964. “Variations in the O^{18}/O^{16} and C^{13}/C^{12} Ratios of Diagenetically Altered Limestones in the Bermuda Islands.” *Journal of Geology* 72 (2): 170–94.

Gussone, Nikolaus, Anne-Sofie C Ahm, Kimberly V Lau, and Harold J Bradbury. 2020. “Calcium Isotopes in Deep Time: Potential and Limitations.” *Chemical Geology*: 119601.

Gussone, Nikolaus, Florian Böhm, Anton Eisenhauer et al. 2005. “Calcium Isotope Fractionation in Calcite and Aragonite.” *Geochimica et Cosmochimica Acta* 69 (18): 4485–94.

Halverson, Galen P, Paul F Hoffman, Daniel P Schrag, Adam C Maloof, and A Hugh N Rice. 2005. “Toward a Neoproterozoic Composite Carbon-Isotope Record.” *GSA Bulletin* 117 (9–10): 1181–207.

Harper, Brandon B, Ángel Puga-Bernabéu, André W Droxler et al.2015. “Mixed Carbonate–Siliciclastic Sedimentation Along the Great Barrier Reef Upper Slope: A Challenge to the Reciprocal Sedimentation Model.” *Journal of Sedimentary Research* 85 (9): 1019–36.

Hayes, John M, Harald Strauss, and Alan J Kaufman. 1999. “The Abundance of ^{13}C in Marine Organic Matter and Isotopic Fractionation in the Global Biogeochemical Cycle of Carbon During the Past 800 Ma.” *Chemical Geology* 161 (1): 103–25.

Haynes, Laura L., and Bärbel Hönisch. 2020. “The Seawater Carbon Inventory at the Paleocene–Eocene Thermal Maximum.” *Proceedings of the National Academy of Sciences* 117 (39): 24088–95.

Henderson, Gideon M, Niall C Slowey, and Geoff A Haddad. 1999. “Fluid Flow Through Carbonate Platforms: Constraints from $^{234}U/^{238}U$ and Cl^- in Bahamas Pore-Waters.” *Earth and Planetary Science Letters* 169 (1–2): 99–111.

Higgins, John A, Clara L Blättler, EA Lundstrom et al. 2018. “Mineralogy, Early Marine Diagenesis, and the Chemistry of Shallow-Water Carbonate Sediments.” *Geochimica et Cosmochimica Acta* 220: 512–34.

Hoffman, Paul F, Dorian S Abbot, Yosef Ashkenazy et al. 2017. "Snowball Earth Climate Dynamics and Cryogenian Geology-Geobiology." *Science Advances* 3 (11): e1600983.

Hoffman, Paul F, and Kelsey G Lamothe. 2019. "Seawater-Buffered Diagenesis, Destruction of Carbon Isotope Excursions, and the Composition of Dic in Neoproterozoic Oceans." *Proceedings of the National Academy of Sciences* 116: 18874–9.

Holland, Steven M. 2020. "The Stratigraphy of Mass Extinctions and Recoveries." *Annual Review of Earth and Planetary Sciences* 48 (1): 75–97.

Hollander, David J, and Judith A McKenzie. 1991. "CO_2 Control on Carbon-Isotope Fractionation During Aqueous Photosynthesis: A Paleo-pCO_2 Barometer." *Geology* 19 (9): 929–32.

Holmden, Chris, RA Creaser, KLSA Muehlenbachs, SA Leslie, and SM Bergstrom. 1998. "Isotopic Evidence for Geochemical Decoupling Between Ancient Epeiric Seas and Bordering Oceans: Implications for Secular Curves." *Geology* 26 (6): 567–70.

Holmden, Chris, K Panchuk, and S C Finney. 2012. "Tightly Coupled Records of ca and C Isotope Changes During the Hirnantian Glaciation Event in an Epeiric Sea Setting." *Geochimica et Cosmochimica Acta* 98: 94–106.

Hovland, Martin, Michael R Talbot, Henning Qvale, Snorre Olaussen, and Lars Aasberg. 1987. "Methane-Related Carbonate Cements in Pockmarks of the North Sea." *Journal of Sedimentary Research* 57 (5): 881–92.

Husson, Jon M, John A Higgins, Adam C Maloof, and Blair Schoene. 2015. "Ca and Mg Isotope Constraints on the Origin of Earth's Deepest $\delta^{13}C$ Excursion." *Geochimica et Cosmochimica Acta* 160: 243–66.

Husson, Jon M, Benjamin J Linzmeier, Kouki Kitajima et al. 2020. "Large Isotopic Variability at the Micron-Scale in 'Shuram' Excursion Carbonates from South Australia." *Earth and Planetary Science Letters* 538: 116211.

Jacobson, Andrew D, and Chris Holmden. 2008. "$\delta^{44}Ca$ Evolution in a Carbonate Aquifer and Its Bearing on the Equilibrium Isotope Fractionation Factor for Calcite." *Earth and Planetary Science Letters* 270 (3): 349–53.

Keigwin, LD, and NJ Shackleton. 1980. "Uppermost Miocene Carbon Isotope Stratigraphy of a Piston Core in the Equatorial Pacific." *Nature* 284 (5757): 613–14.

Keith, ML, and JN Weber. 1964. "Carbon and Oxygen Isotopic Composition of Selected Limestones and Fossils." *Geochimica et Cosmochimica Acta* 28 (10–11): 1787–816.

Khadka, Mitra B., Jonathan B. Martin, and Jin Jin. 2014. "Transport of Dissolved Carbon and CO_2 Degassing from a River System in a Mixed Silicate and Carbonate Catchment." *Journal of Hydrology* 513: 391–402.

King, Arthur S, and Raymond T Birge. 1929. "An Isotope of Carbon, Mass 13." *Nature* 124 (3117): 127–27.

Knauth, L Paul, and Martin J Kennedy. 2009. "The Late Precambrian Greening of the Earth." *Nature* 460 (7256): 728–32.

Knoll, AH, JM Hayes, AJ Kaufman, K Swett, and IB Lambert. 1986. "Secular Variation in Carbon Isotope Ratios from Upper Proterozoic Successions of Svalbard and East Greenland." *Nature* 321 (6073): 832–8.

Kohout, FA. 1965. "A Hypothesis Concerning Cyclic Flow of Salt Water Related to Geothermal Heating in the Floridan Aquifer." *New York Academy of Sciences Transactions* 28: 249–71.

Komar, N, and RE Zeebe. 2016. "Calcium and Calcium Isotope Changes During Carbon Cycle Perturbations at the End-Permian: End-Permian Calcium Cycle." *Paleoceanography* 31 (1): 115–30.

Kump, Lee R, and Michael A Arthur. 1999. "Interpreting Carbon-Isotope Excursions: Carbonates and Organic Matter." *Chemical Geology* 161 (1–3): 181–98.

Kump, Lee R, M A Arthur, M E Patzkowsky et al. 1999. "A Weathering Hypothesis for Glaciation at High Atmospheric pCO_2 During the Late Ordovician." *Palaeogeography, Palaeoclimatology, Palaeoecology* 152 (1): 173–87.

Laakso, Thomas A, and Daniel P Schrag. 2020. "The Role of Authigenic Carbonate in Neoproterozoic Carbon Isotope Excursions." *Earth and Planetary Science Letters* 549: 116534.

Lazar, Boaz, and Jonathan Erez. 1992. "Carbon Geochemistry of Marine-Derived Brines: I. $\delta^{13}C$ Depletions Due to Intense Photosynthesis." *Geochimica et Cosmochimica Acta* 56 (1): 335–45.

Li, Juan, Xiumian Hu, Eduardo Garzanti, and Marcelle BouDagher-Fadel. 2021. "Climate-Driven Hydrological Change and Carbonate Platform Demise Induced by the Paleocene–Eocene Thermal Maximum (Southern Pyrenees)." *Palaeogeography, Palaeoclimatology, Palaeoecology*: 110250.

Li, Juan, Xiumian Hu, James C Zachos, Eduardo Garzanti, and Marcelle BouDagher-Fadel. 2020. "Sea Level, Biotic and Carbon-Isotope Response to the Paleocene–Eocene Thermal Maximum in Tibetan Himalayan Platform Carbonates." *Global and Planetary Change* 194: 103316.

Loutit, Tom S, and James P Kennett. 1979. "Application of Carbon Isotope Stratigraphy to Late Miocene Shallow Marine Sediments, New Zealand." *Science* 204 (4398): 1196–9.

Lowenstam, Heinz A, and Samuel Epstein. 1957. "On the Origin of Sedimentary Aragonite Needles of the Great Bahama Bank." *Journal of Geology* 65 (4): 364–75.

Lynch-Stieglitz, Jean, Thomas F Stocker, Wallace S Broecker, and Richard G Fairbanks. 1995. "The Influence of Air-Sea Exchange on the Isotopic Composition of Oceanic Carbon: Observations and Modeling." *Global Biogeochemical Cycles* 9 (4): 653–65.

Maher, D.T., I.R. Santos, L. Golsby-Smith, J. Gleeson, and B.D. Eyre. 2013. "Groundwater-Derived Dissolved Inorganic and Organic Carbon Exports from a Mangrove Tidal Creek: The Missing Mangrove Carbon Sink?" *Limnology and Oceanography* 58 (2): 475–88.

McInerney, Francesca A., and Scott L. Wing. 2011. "The Paleocene-Eocene Thermal Maximum: A Perturbation of Carbon Cycle, Climate, and Biosphere with Implications for the Future." *Annual Review of Earth and Planetary Sciences* 39 (1): 489–516.

Melim, Leslie A, Peter K Swart, and Robert G Maliva. 1995. "Meteoric-Like Fabrics Forming in Marine Waters: Implications for the Use of Petrography to Identify Diagenetic Environments." *Geology*, 4.

Melim, Leslie A, Peter K. Swart, and Robert G Maliva. 2001. "Meteoric and Marine Burial Diagenesis in the Subsurface of Great Bahama Bank." *SEPM Special Publication*, 25.

Melim, L.A, H Westphal, P.K Swart, G. P Eberli, and A Munnecke. 2002. "Questioning Carbonate Diagenetic Paradigms: Evidence from the Neogene of the Bahamas." *Marine Geology* 185 (1): 27–53.

Murray, Sean T., John A. Higgins, Chris Holmden, Chaojin Lu, and Peter K. Swart. 2021. "Geochemical Fingerprints of Dolomitization in Bahamian Carbonates: Evidence from Sulphur, Calcium, Magnesium and Clumped Isotopes." *Sedimentology* 68 (1): 1–29.

Naehr, T.H., N.M. Rodriguez, G. Bohrmann, C.K. Paull, and R. Botz. 2000. "Methane-Derived Authigenic Carbonates Associated with Gas Hydrate Decomposition and Fluid Venting Above the Blake Ridge Diapir." In *Proceedings of the Ocean Drilling Program, 164 Scientific Results*. Vol. 164. Proceedings of the Ocean Drilling Program. Ocean Drilling Program.

Nier, Alfred O, and Earl A Gulbransen. 1939. "Variations in the Relative Abundance of the Carbon Isotopes." *Journal of the American Chemical Society* 61 (3). ACS Publications: 697–98.

Oehlert, Amanda M, Kathryn A Lamb-Wozniak, Quinn B Devlin et al.2012 "The Stable Carbon Isotopic Composition of Organic Material in Platform Derived Sediments: Implications for Reconstructing the Global Carbon Cycle." *Sedimentology* 59 (1). Wiley Online Library: 319–35.

Oehlert, Amanda M, and Peter K Swart. 2014. "Interpreting Carbonate and Organic Carbon Isotope Covariance in the Sedimentary Record." *Nature Communications* 5 (1). Nature Publishing Group: 1–7.

Oehlert, Amanda M., and Peter K. Swart. 2019. "Rolling Window Regression of $\delta^{13}C$ and $\delta^{18}O$ Values in Carbonate Sediments: Implications for Source and Diagenesis." *Depositional Record* 5 (3): 613–30.

Opdyke, Bradley N., and Bruce H. Wilkinson. 1988. "Surface Area Control of Shallow Cratonic to Deep Marine Carbonate Accumulation." *Paleoceanography* 3 (6): 685–703.

Panchuk, Karla M., Chris E. Holmden, and Stephen A. Leslie. 2006. "Local Controls on Carbon Cycling in the Ordovician Midcontinent Region of North America, with Implications for Carbon Isotope Secular Curves." *Journal of Sedimentary Research* 76 (2): 200–11.

Pancost, Richard D., Katherine H. Freeman, and Mark E. Patzkowsky. 1999. "Organic-Matter Source Variation and the Expression of a Late Middle Ordovician Carbon Isotope Excursion." *Geology* 27 (11): 1015–18.

Patterson, William P, and Lynn M Walter. 1994. "Depletion of ^{13}C in Seawater ΣCO_2 on Modern Carbonate Platforms: Significance for the Carbon Isotopic Record of Carbonates." *Geology* 22 (10). Geological Society of America: 885–8.

Popp, Brian N, Edward A Laws, Robert R Bidigare et al. 1998. "Effect of Phytoplankton Cell Geometry on Carbon Isotopic Fractionation." *Geochimica et Cosmochimica Acta* 62 (1). Elsevier: 69–77.

Prahl, F. G, J. R Ertel, M. A Goni, M. A Sparrow, and B Eversmeyer. 1994. "Terrestrial Organic Carbon Contributions to Sediments on the Washington Margin." *Geochimica et Cosmochimica Acta* 58 (14): 3035–48.

Richardson, Christina M., Henrietta Dulai, Brian N. Popp, Kathleen Ruttenberg, and Joseph K. Fackrell. 2017. "Submarine Groundwater Discharge Drives Biogeochemistry in Two Hawaiian Reefs." *Limnology and Oceanography* 62 (S1): S348–S363.

Rodriguez Blanco, Leticia, Gregor P. Eberli, Ralf J. Weger et al. 2020. "Periplatform Ooze in a Mixed Siliciclastic-Carbonate System – Vaca Muerta Formation, Argentina." *Sedimentary Geology* 396: 105521.

Röhl, Ursula, Thomas Westerhold, Timothy J. Bralower, and James C. Zachos. 2007. "On the Duration of the Paleocene-Eocene Thermal Maximum (PETM)." *Geochemistry, Geophysics, Geosystems* 8 (12).

Romanek, Christopher S, Ethan L Grossman, and John W Morse. 1992. "Carbon Isotopic Fractionation in Synthetic Aragonite and Calcite: Effects of Temperature and Precipitation Rate." *Geochimica et Cosmochimica Acta* 56 (1). Elsevier: 419–30.

Ronov, AB, VE Khain, AN Balukhovsky, and KB Seslavinsky. 1980. "Quantitative Analysis of Phanerozoic Sedimentation." *Sedimentary Geology* 25 (4). Elsevier: 311–25.

Rooney, Alan D, Marjorie D Cantine, Kristin D Bergmann et al. 2020. "Calibrating the Coevolution of Ediacaran Life and Environment." *Proceedings of the National Academy of Sciences* 117 (29). National Academy of Sciences: 16824–30.

Rothman, D.H., J.M. Hayes, and R.E. Summons. 2003. "Dynamics of the Neoproterozoic Carbon Cycle." *Proceedings of the National Academy of Sciences* 100 (14): 8124–9.

Saltzman, Matthew R. 2005. "Phosphorus, Nitrogen, and the Redox Evolution of the Paleozoic Oceans." *Geology* 33 (7). Geological Society of America: 573–6.

Schidlowski, Manfred, Rudolf Eichmann, and Christian E Junge. 1975. "Precambrian Sedimentary Carbonates: Carbon and Oxygen Isotope Geochemistry and Implications for the Terrestrial Oxygen Budget." *Precambrian Research* 2 (1). Elsevier: 1–69.

Schlager, Wolfgang, John JG Reijmer, and AW Droxler. 1994. "Highstand Shedding of Carbonate Platforms." *Journal of Sedimentary Research* 64 (3b). SEPM Society for Sedimentary Geology: 270–81.

Schoene, Blair. 2014. "U–Th–Pb Geochronology." In *Treatise on Geochemistry*, eds. Heinrich D. Holland and Karl K. Turekian, 2nd ed., 341–78. Elsevier.

Scholle, Peter A, and Michael A Arthur. 1980. "Carbon Isotope Fluctuations in Cretaceous Pelagic Limestones: Potential Stratigraphic and Petroleum Exploration Tool." *AAPG Bulletin* 64 (1). American Association of Petroleum Geologists (AAPG): 67–87.

Schrag, Daniel P, Robert A Berner, Paul F Hoffman, and G.P. Halverson. 2002. "On the Initiation of a Snowball Earth." *Geochemistry, Geophysics, and Geosystems* 300.

Schrag, Daniel P, John A Higgins, Francis A Macdonald, and David T Johnston. 2013. "Authigenic Carbonate and the History of the Global Carbon Cycle." *Science* 339 (6119). American Association for the Advancement of Science: 540–3.

Skulan, Joseph, Donald J. DePaolo, and Thomas L. Owens. 1997. "Biological Control of Calcium Isotopic Abundances in the Global Calcium Cycle." *Geochimica et Cosmochimica Acta* 61 (12): 2505–10.

Sluijs, Appy, Henk Brinkhuis, Erica M. Crouch et al. 2008. "Eustatic Variations During the Paleocene-Eocene Greenhouse World." *Paleoceanography* 23 (4).

Staudigel, Philip T., and Peter K. Swart. 2019. "A Diagenetic Origin for Isotopic Variability of Sediments Deposited on the Margin of Great Bahama Bank, Insights from Clumped Isotopes." *Geochimica et Cosmochimica Acta* 258: 97–119.

Swanson-Hysell, Nicholas L, Adam C Maloof, Daniel J Condon et al. 2015. "Stratigraphy and Geochronology of the Tambien Group, Ethiopia: Evidence for Globally Synchronous Carbon Isotope Change in the Neoproterozoic." *Geology* 43 (4). Geological Society of America: 323–6.

Swart, Peter K. 2008. "Global Synchronous Changes in the Carbon Isotopic Composition of Carbonate Sediments Unrelated to Changes in the Global Carbon Cycle." *Proceedings of the National Academy of Sciences* 105 (37). National Academy of Sciences: 13741–5.

Swart, Peter K. 2015. "The Geochemistry of Carbonate Diagenesis: The Past, Present and Future." *Sedimentology* 62 (5). Wiley Online Library: 1233–1304.

Swart, Peter K, and Gregor P Eberli. 2005. "The nature of the $\delta^{13}C$ of periplatform sediments: Implications for stratigraphy and the global carbon cycle." *Sedimentary Geology* 175 (1–4). Elsevier: 115–29.

Tang, Jianwu, Martin Dietzel, Florian Böhm, Stephan J Köhler, and Anton Eisenhauer. 2008. "Sr^{2+}/Ca^{2+} and $^{44}Ca/^{40}Ca$ Fractionation During Inorganic Calcite Formation: II. Ca Isotopes." *Geochimica et Cosmochimica Acta* 72 (15). Elsevier: 3733–45.

Tziperman, E., I. Halevy, D. T. Johnston, A. H. Knoll, and D. P. Schrag. 2011. "Biologically Induced Initiation of Neoproterozoic Snowball-Earth Events." *Proceedings of the National Academy of Sciences* 108 (37): 15091–6.

Urey, Harold C. 1947. "The Thermodynamic Properties of Isotopic Substances." *Journal of the Chemical Society (Resumed)*. Royal Society of Chemistry, 562–81.

Urey, Harold C, A.H.W. Aten Jr., and Albert S Keston. 1936. "A Concentration of the Carbon Isotope." *Journal of Chemical Physics* 4 (9). American Institute of Physics: 622–3.

Vahrenkamp, Volker C, Peter K Swart, and Joaquin Ruiz. 1991. "Episodic Dolomitization of Late Cenozoic Carbonates in the Bahamas; Evidence from Strontium Isotopes." *Journal of Sedimentary Research* 61 (6). SEPM Society for Sedimentary Geology: 1002–14.

Walker, James CG, PB Hays, and James F Kasting. 1981. "A Negative Feedback Mechanism for the Long-Term Stabilization of Earth's Surface Temperature." *Journal of Geophysical Research: Oceans* 86 (C10). Wiley Online Library: 9776–82.

Wanninkhof, Rik. 1985. "Kinetic Fractionation of the Carbon Isotopes ^{13}C and ^{12}C During Transfer of CO_2 from Air to Seawater." *Tellus B* 37B (3): 128–35.

Westerhold, Thomas, Norbert Marwan, Anna Joy Drury et al.2020. "An Astronomically Dated Record of Earth's Climate and Its Predictability over the Last 66 Million Years." *Science* 369 (6509). American Association for the Advancement of Science: 1383–7.

Westerhold, Thomas, Ursula Röhl, Roy H. Wilkens et al. 2018. "Synchronizing Early Eocene Deep-Sea and Continental Records – Cyclostratigraphic Age Models for the Bighorn Basin Coring Project Drill Cores." *Climate of the Past* 14 (3): 303–19.

Zachos, James C., Steven M Bohaty, Cedric M John et al. 2007. "The Palaeocene–Eocene Carbon Isotope Excursion: Constraints from Individual Shell Planktonic Foraminifer Records." *Philosophical Transactions of the Royal Society A: Mathematical, Physical and Engineering Sciences* 365 (1856): 1829–42.

Zachos, James C., Ursula Röhl, Stephen A. Schellenberg et al. 2005. "Rapid Acidification of the Ocean During the Paleocene-Eocene Thermal Maximum." *Science* 308 (5728): 1611–5.

Zachos, James C., Michael W. Wara, Steven Bohaty et al. 2003. "A Transient Rise in Tropical Sea Surface Temperature During the Paleocene-Eocene Thermal Maximum." *Science* 302 (5650): 1551–4.

Zeebe, Richard E., Jelle Bijma, and Dieter A. Wolf-Gladrow. 1999. "A Diffusion-Reaction Model of Carbon Isotope Fractionation in Foraminifera." *Marine Chemistry* 64 (3): 199–227.

Zeebe, Richard E., and Lucas J. Lourens. 2019. "Solar System Chaos and the Paleocene–Eocene Boundary Age Constrained by Geology and Astronomy." *Science* 365 (6456): 926–9.

Zhou, Xiaoli, Ellen Thomas, Rosalind E. M. Rickaby, Arne M. E. Winguth, and Zunli Lu. 2014. "I/Ca Evidence for Upper Ocean Deoxygenation During the PETM." *Paleoceanography* 29 (10): 964–75.

Acknowledgments

We thank Tim Lyons for editorial handing. We are grateful to both Benjamin Gill and one anonymous reviewer for careful and constructive comments that significantly improved the manuscript. This work was supported by a grant from the Simons Foundation (SCOL 611878, ASCA).

Cambridge Elements

Geochemical Tracers in Earth System Science

About the Series

This innovative series provides authoritative, concise overviews of the many novel isotope and elemental systems that can be used as 'proxies' or 'geochemical tracers' to reconstruct past environments over thousands to millions to billions of years – from the evolving chemistry of the atmosphere and oceans to their cause-and-effect relationships with life.

Covering a wide variety of geochemical tracers, the series reviews each method in terms of the geochemical underpinnings, the promises and pitfalls, and the 'state-of-the-art' and future prospects, providing a dynamic reference resource for graduate students, researchers and scientists in geochemistry, astrobiology, paleontology, paleoceanography and paleoclimatology.

The short, timely, broadly accessible papers provide much-needed primers for a wide audience – highlighting the cutting-edge of both new and established proxies as applied to diverse questions about Earth system evolution over wide-ranging time scales.

Cambridge Elements

Geochemical Tracers in Earth System Science

Elements in the Series

A full series listing is available at: www.cambridge.org/EESS

Printed in the United States
by Baker & Taylor Publisher Services